CW01648183

Hume's discussion of the idea of space in his *Treatise on Human Nature* is fundamental to an understanding of his treatment of such central issues as the existence of external objects, the unity of the self, the relation between certainty and belief, and abstract ideas. Marina Frasca-Spada's rich and original study examines this difficult part of Hume's philosophical writings and connects it to eighteenth-century works in natural philosophy, mathematics, and literature. Focussing on Hume's discussions of the infinite divisibility of extension, the origin of the idea of space, geometry, and the notion of a vacuum, she shows that the central questions of Hume's 'science of human nature' – what does the 'science of human nature' reveal about the mind and its operations? What is experience? – underlie all of these discussions. Her analysis points the way to a reassessment of the central current interpretative problems in Hume studies.

SPACE AND THE SELF IN HUME'S *TREATISE*

SPACE AND THE SELF IN HUME'S *TREATISE*

MARINA FRASCA-SPADA
University of Cambridge

CAMBRIDGE
UNIVERSITY PRESS

PUBLISHED BY THE PRESS SYNDICATE OF THE UNIVERSITY OF CAMBRIDGE
The Pitt Building, Trumpington Street, Cambridge, United Kingdom

CAMBRIDGE UNIVERSITY PRESS
The Edinburgh Building, Cambridge CB2 2RU, UK
40 West 20th Street, New York NY 10011–4211, USA
477 Williamstown Road, Port Melbourne, VIC 3207, Australia
Ruiz de Alarcón 13, 28014 Madrid, Spain
Dock House, The Waterfront, Cape Town 8001, South Africa

http://www.cambridge.org

First published 1998
First paperback edition 2002

Typeface Baskerville 11/12.5 pt.

A catalogue record for this book is available from the British Library

Library of Congress Cataloguing in Publication data
Frasca-Spada, Marina.
Space and the self in Hume's *Treatise* / Marina Frasca-Spada.
p. cm.
Papers presented on Hume and the idea of space at several places
between 1992 and 1995.
Includes bibliographical references and index.
ISBN 0 521 62091 0 (hardback)
1. Hume, David, 1711–1776. Treatise of human nature. 2. Space
and time. 3. Hume, David, 1711–1776 – Contributions in notion of
space. 4. Hume, David, 1711–1776 – Contributions in doctrine of the
self in philosophy. 5. Self (Philosophy) I. Title.
B1489.F73 1998
128–dc21 98-25535 CIP

ISBN 0 521 62091 0 hardback
ISBN 0 521 89162 0 paperback

To the memory of Raniero Sabarini, who introduced me to philosophy, and to Boris and Romilly Jardine

Contents

Acknowledgements

I have incurred many and various debts in the course of the many years of puzzlement with Book 1, Part 2 of Hume's *Treatise* that led to the work that follows. First and foremost there is the debt to Peter Lipton, whose teaching has been inspiring, and whose support essential.

I thank Tullio Gregory for focussing my attention on Hume and the idea of space, and I feel deeply indebted to Paolo Casini for his patient and kind interest in my philosophical perplexities at an early stage of my work. My special thanks to John Yolton for his extensive and detailed comments and for his generous encouragement, and to Nick Jardine, who has read and discussed all my work with me. I have received sympathetic and stimulating feedback from two anonymous referees for Cambridge University Press, exceptionally prompt and encouraging responses from the Press editor, Hilary Gaskin, and excellent and expert advice from Gillian Maude, the Press copyeditor. I am also grateful to Janet Dudley for her assistance in preparing the index.

For reading and commenting on earlier versions of this work, I thank Michael Ayers, Annette Baier, Edward Craig, Andrew Cunningham, Serafina Cuomo, Alessandra D'Acconti, Constance Dilworth, Martin Kusch, Peter Jones, Donald Livingston, Sandro Petruccioli, Simon Schaffer, Emma Spary.

For stimulating (if sometimes nettling) discussions on the meaning of life, history writing, philosophy, and David Hume, I am grateful to Richard Ashcroft, Hazel Benyon, Cristina Chimisso, Lucia Dacome, Maria Elena Di Stefano, Paolo Freguglia, Neil Gascoigne, Enrico Giusti, Adrian Johns, Sachiko Kusukawa, Bruno Morcavallo, Sue Morgan, Beppo Occhialini, Paul Ries, the members of the Cambridge Historiography Group, the members of the Epistemology Reading Group, the attendants at the 1992 Hume Society Annual Conference in Nantes, in particular Tim Engström, Rebecca Kukla, Peter Millikan, and Patrick Murray, and the participants at the 1996 Hume

Society Annual Conference in Nottingham, in particular Lorne Falkenstein.

I owe much to the Principal and the Fellows of Newnham College, Cambridge, where I held a Research Fellowship; they were an inexhaustible source of advice, sympathy, and encouragement at every stage of the writing and rewriting.

I am also grateful to Rebecca Bertoloni Meli, Nancy Sempers, and Linda Washington, of the Whipple Library, and to the staff of the Cambridge University Library, especially of the Rare Books Room.

My work has been funded by, among others, the British Academy, St John's College, and Newnham College; but it would have been unthinkable without the support (financial and otherwise) of my parents, Laura Dei and Mario Frasca-Spada. To them go my deepest thanks.

A version of chapter 1 was published as a two-part article in *The British Journal for the History of Philosophy* (5/2 (1997), and 6/1 (1998)); I am grateful to the Editor, John Rogers, for permission to make use of this material.

Abbreviations

T	David Hume, *A Treatise of Human Nature*, ed. with analytical index by L. A. Selby-Bigge, 2nd edition with text revised and variant readings by P. H. Nidditch, Oxford: Clarendon Press, 1978.
E	David Hume, *Enquiries concerning Human Understanding and concerning the Principles of Morals*, ed. with introduction, comparative table of contents, and analytical index by L. A. Selby-Bigge, 3rd edition, with text revised and notes by P. H. Nidditch, Oxford: Clarendon Press, 1975.
A	*An Abstract of a Book Lately Published; Entituled, a Treatise of Human Nature, &c. Wherein The Chief Argument of that Book is farther illustrated and explained*, in L. A. Selby-Bigge (ed.), *Hume's Treatise*.
Greig 1/2	J. Y. T. Greig (ed.), *Letters of David Hume*, 2 vols., Oxford: Clarendon Press, 1932.

Introduction

> He is not likely to know what is to be done unless he lives in what is not merely the present, but the present moment of the past, unless he is conscious, not of what is dead, but of what is already living.
>
> (T. S. Eliot, 'Tradition and the individual talent', *Selected Essays*, 3rd edition, London 1951, p. 22)

The work that follows is an interpretation of Hume's treatment of the idea of space. Let me start *in medias res* by mentioning some well-known facts. This is a part of Hume's philosophy which appears with some prominence in his *Treatise of Human Nature* – it is the subject of Book 1, Part 2 – and hardly at all in the *Enquiry Concerning Human Understanding*, and which has often been dismissed as weak and unsatisfactory; its absence has even been seen as a sign of Hume's greater maturity of thought in the *Enquiry*.[1] One of the assumptions of the present work is, indeed, that the case of the idea of space typifies the difference between the *Treatise* and the *Enquiry*. Accordingly I shall show how closely the treatment of the idea of space is bound up with other themes of Hume's theory of the understanding that are central in the *Treatise* and that disappear or whose discussion is radically abridged and simplified in the *Enquiry*: for instance, the problem of the existence of external objects, the problem of the self, the problem of abstract ideas. From this point of view, the treatment of the idea of space appears as crucial in the definition of the philosophical substance of the *Treatise*. So, the question of why I have decided to examine Hume's theory of the understanding through a detailed analysis of the treatment of the idea of space, may, in fact, be generalised: why have I decided to privilege the *Treatise* – especially since Hume himself was famously so explicit in advising his readers to do the opposite?

A first answer is that I find the *Treatise* more fun than the *Enquiries* – in

[1] For instance, see J. Noxon, *Hume's Philosophical Development*, Oxford 1973, pp. 114ff., and A. Flew, *Hume's Philosophy of Belief. A Study of His First Inquiry*, London 1961, pp. 61ff.

other words, it is in large part a matter of taste. Yet it may be of some interest to outline what is involved in concentrating on material peculiar to the *Treatise*, while showing how the idea of space is central to any such reading. This is what I do in the *Intermezzo*, where I tell of the reactions to, and appropriations of, the *Treatise* by Hume himself and by some other contemporary and more recent readers, organising the story around the general problems of authorial intention and contemporary reception.

But now it is expedient to give a proper preview of the work. To start with, my discussion follows the order of Book 1, Part 2 of the *Treatise*. Chapter 1 examines Hume's rejection of the 'doctrine of infinite divisibility', with which he opens his treatment of the idea of space (sects. 1–2). The focus of this chapter is on how the mind establishes the relation between mental and real objects: I show that the key feature of the overall argument of sections 1–2 is Hume's way of articulating the passage from the existence of indivisible impressions and ideas to the impossibility of the infinite divisibility of real extension. Chapter 2 is on the origin of the idea of space (sect. 3) from the 'disposition' or 'manner of appearance' of the indivisible impressions or perceptual data. An apparent exception to the first principle that all our ideas are copies of impressions, the notion of 'manner of appearance' is, I suggest, related to belief as 'manner of conception', and may be seen as expressing an original contribution of the mind to experience. With this, my examination of Hume's direct presentation of his doctrine is over. The following two sections of the *Treatise* contain what he calls 'Objections answer'd'; and chapters 3 and 4, which follow the *Intermezzo* on reading and interpretation, are devoted to them. These chapters expand on themes introduced or intimated in chapters 1 and 2. Chapter 3 addresses the relation between knowledge and belief and the classic Kantian question of the Humean a priori by focussing on Hume's treatment of geometry as a source of possible objections against the theory of perceptual indivisibles (sect. 4), and his location of geometry in an ambiguous position between knowledge and probability (Part 3, sect. 1). At the end of this chapter, I consider Hume's treatment of curiosity and the love of truth in the conclusion of the book on passions (Book 2, Part 3, sect. 10). Chapter 4 is devoted to Hume's discussion of the idea of empty space (sect. 5). I show how, according to Hume, we only have a pseudo-idea of empty space and a vacuum which is associated with, and brought about by, our use of the terms 'empty space' and 'vacuum'. In the light of this, I suggest that the formation of the very idea of space from the 'dispositions' or 'manners of appearance' of perceptual objects involves a reflexive act of the mind similar to those responsible for the formation of abstract ideas by means of distinctions of reason. I conclude the chapter with remarks on the role of sociability and conversation in Hume's theory of language and the understanding.

These chapters are relatively independent, if obviously associated studies, each with a definite scope and with a precise task. By the end of the final chapter, however, it will be evident that one fundamental question underlies each of Hume's sections on the idea of space, making them, as well as my readings, variations on the same theme: what is a mind, and what and how can it know about itself?

The precise task I set for my four chapters is, in each case, that of interpreting an exemplary obscure or puzzling passage from the relevant section of Hume's *Treatise*: four passages – on a grain of sand (*T*/27), on the impressions offered by the surface of the table (*T*/34), on the standard of equality and the definition of a straight line (*T*/51), and on the use of the word 'vacuum' (*T*/64) – define the fields of investigation of my four studies; and the interpretation is the occasion for a reconstruction of the principles involved. My reconstructions are not meant, of course, to say the final word on those principles, but simply to reveal aspects of them which would perhaps remain in the dark in a more systematic analysis. So my writing about Hume's philosophy does not attempt to tidy it up, and is deliberately regressive. It is a feature of this procedure that interpretation of a passage starts with the often laborious accumulation of material – through examination of context, reconstruction of sources and contemporary usage, close reading and comparison of parallel passages – before a reconstructive argument can be presented. This technique draws, to some extent, on the traditions of commentary and argumentative reconstruction. Some philosophical texts, for example Kant's *Critique of Pure Reason*, invite, or even demand this approach;[2] because of its local clarity and

[2] See D. Henrich, 'Identität und Objektivität. Eine Untersuchung über Kants tranzendentale Deduktion', *Sitzungsberichte der Heidelberger Akademie der Wissenschaften. Philosophisch-historische Klasse*, 1976, pp. 1–112, esp. pp. 1ff., on the difficulties of Kantian interpretation. The ideal set by Henrich is, in my view, revealing: 'One can only follow the texts of such an author really closely if one is sure of always being able to decide where one is staying within the bounds of his point of view, and where one begins to substitute for it other, though related and possibly also more convincing, ones. Kant's basic ideas will only come to be understood when one is able to bring the force of analysis of arguments fully into the analysis of texts. If that is not achieved, philosophical investigations will result which are related to Kantian thought and orchestrated with Kantian quotations; but Kant's historical attempt at a theory will remain in the twilight from which no one has yet been able to draw it clear' (p. 14).

apparent smoothness (at least in the *Enquiries*), Hume's prose does not. But many will agree, I think, that we cannot regard the text of the *Treatise* as a transparent vehicle of philosophical ideas, and it is my conviction that interpretative effort focussed on especially irksome parts may cast a sharper or more interesting light on the *Treatise* as a whole. This is why, rather than starting with a chapter on the copy principle and concluding with, say, a chapter on the meaning of Hume's 'science of human nature', this interpretation contains, among other things, a series of sketches of the copy principle and of the 'science of human nature' from different, sometimes unusual, perspectives.

So my investigations do not single out certain principles as the essential ones from which to derive consequences in order to build up a Humean philosophical system. On the other hand, my reading of Hume's treatment of the idea of space does claim to be, in fact, a reading of Hume's theory of human understanding concentrated on his treatment of the idea of space. This claim may naturally suggest to the reader some more general questions – what kind of Humean philosophy is presupposed by and constituted in this reading? Is it naturalistic or sceptical? Answering such questions is very difficult – indeed, making it clear that answering such questions is very difficult, and showing why, are among my aims here. As will become evident, I am reluctant to use such summarising terms as 'naturalism', 'scepticism', or 'realism' because they are, in my opinion, unhelpful – inadequate to decribe the complexities of the real thing. 'Scepticism', 'naturalism', and 'realism' are not the only summarising terms I am reluctant to use. I would also feel uneasy repeating once more that, for example, Hume has a psychological rather than a logical ideal of certainty, or a psychological rather than a philosophical interest in the problem of induction, and so on. Nor do I think that calling Hume's 'science of human nature' a 'psychology', and contrasting it with a 'philosophy' or a 'logic' would be a good interpretative move. That post-Humean categories and expectations, including our own, focus and direct our readings is inevitable; their unreflective use is a different matter. An ancient text such as Hume's *Treatise* is a very delicate object – or, to put it more accurately, we, as present-day readers of Hume's *Treatise* and the like, are very delicate objects: our balance may be easily upset by uncautiously using our, rather than its own, terminology, and we risk reading into a marvellous eighteenth-century book a second-rate twentieth-century one.

The part of Hume's philosophical agenda that I shall reconstruct, whilst doubtless in some loose sense 'naturalistic' or 'pragmatist', has little in common with that of the *present-day* sceptic, or realist, or even naturalist and pragmatist. Given the liveliness of contemporary debate on these issues one point, however, must be made clear. This study contains an interpretation of Hume's theory of the understanding based on a close reading of the treatment of the idea of space, and a reconstruction of its connection with the central doctrines of the *Treatise*. Of course, I believe this interpretation to be new, and also to be legitimate; and I think that the methodology I have used is reasonable and fruitful. But with this I do not intend to suggest that my reading of Hume is the only defensible one, or that my approach is the only one leading to an attractive and convincing interpretation. This would be simply incompatible with the approach I have adopted. Provided they are based on a decent awareness of the past categories and values operating in the past text, approaches more informed by current philosophical (or other) agendas may result in very good readings indeed.[3]

But the prescription to avoid unreflective overriding, or indeed unreflective adoption, of the relevant past categories leaves plenty of scope for divergent interpretations. The attitude to the text that I am advocating requires that, within these limits, we regard Hume's *Treatise* in the same way that many literary critics consider classics: that is, as by their very definition, open to different – differently oriented and ever new – interpretations.[4] This move is clearly appropriate to the historical fortunes of the book we are considering. Seen sometimes as a founding father, sometimes as an exemplary erroneous thinker,[5] the author of the *Treatise* is always in the philosophical pantheon: of analytic philosophers, still paying their typically grudging respects to the first who posed the problems of causation and induction;[6] of phenomenologists, ready to acknowledge the similarity of their reductions with

[3] For example, S. Blackburn, 'Hume on thick connexions' (*Essays in Quasi-Realism*, Oxford 1993, chapter 5, pp. 94–107), offers an exceptionally stimulating 'quasi-realist' Hume by freely improving or criticising Hume's arguments whenever he thinks it useful or necessary, but without for a moment forgetting to be fair and respectful to the text and its pastness. A very good reading of Hume informed by a current socio-political agenda is in T. Eagleton, *The Ideology of the Aesthetic*, Oxford 1990, chapter 2: 'The law of the heart', esp. pp. 45–52.

[4] See F. Kermode, *The Classic*, Cambridge (Mass.) 1983, p. 44: 'It seems that on a just view of the matter the books we call classics possess intrinsic qualities that endure, but possess also an openness to accommodation which keeps them alive under endlessly varying dispositions.'

[5] For Hume as a very bad thing see, for example, C. D. Broad, 'Hume's theory of space', *Proceedings of the British Academy* 47, 1961, pp. 161–76 (and references there).

[6] See, for example, J. Passmore, *Hume's Intentions*, 3rd edition, London 1980, and J. Bennett, *Locke Berkeley Hume: Central Themes*, Oxford 1971. For the anti-canon of analytic philosophy see J. Rée, *Philosophical Tales*, London and New York 1987, pp. 42ff.

Hume's examinations of experience;[7] of positivists, who have seen in his writings the first appearance of their own anti-metaphysical doctrines;[8] as well as of deconstructionists and post-moderns, who, without much effort, have appropriated both his concern with style and writing, and what he calls his 'moderate scepticism' and 'philosophy in a careless manner'.[9] There can be really no doubt about the status of Hume as a canonical philosopher with a vengeance.[10]

This brings me to another important point. My reading of Book 1, Part 2 of the *Treatise* is both closely textual and contextual. But the types of context to which I shall appeal in my readings will perhaps look somewhat idiosyncratic, having reference more to readership and reception, than to influence and authorial intentions. For example, I shall deliberately avoid comparing Hume's tenets with Locke's and Berkeley's. There are several other reasons for this: in particular a certain dissatisfaction with the standard story of the British empiricists, which, at any rate, has been carefully explored by many others. Also, every line of such authors as Locke and Berkeley has been subjected to nearly three centuries of comments, interpretations, refutations, imitations, developments – their texts are so charged with meanings of all sorts and origins, that there is room for doubt about their efficacy in sorting out interpretative problems in the text of Hume. In my contextual readings, I shall set out to make sense of Hume's pages by reconstructing the climate of opinion to which they belonged, rather than by establishing a system of influences among authors of philosophy classics. For example, I shall consider Hume's relation to the 'way of ideas',

[7] See A. Reinach, 'Kants Auffassung des humeschen Problem', *Zeitschrift für Philosophie und philosophische Kritik*, 141 (1908), pp. 176–209 (English transl. by J. N. Mohanty, in K. R. Merrill and R. W. Shahan (eds.), *David Hume: Many-sided Genius*, Norman 1976, pp. 161–88), and C. V. Salmon, 'The central problem of David Hume's philosophy: an essay toward a phenomenological interpretation of the first book of the *Treatise of Human Nature*', *Jahrbuch für Philosophie und phänomenologische Forschung* X, 1929, 299–449. See the discussion of G. Davie, *A Passion for Ideas: Essays on the Scottish Enlightenment*, vol. 2, Edinburgh 1994, chapters 7 and 8; interesting material in D. W. Livingston, *Hume's Philosophy of Common Life*, Chicago 1984, pp. 48ff.

[8] For example, see H. Reichenbach, *The Rise of Scientific Philosophy*, Berkeley 1951, and A. J. Ayer, *Language, Truth and Logic*, 2nd edition, London 1946 (as well as his *Hume*, Oxford 1980), and such literature as F. Zabeeh, *Hume Precursor of Modern Empiricism*, The Hague 1973, and D. F. Pears (ed.), *David Hume: A Symposium*, London 1963.

[9] See, for instance, Z. Parusnikova, 'Against the spirit of foundations: post-modernism and David Hume', *Hume Studies* 19 (1), 1993, pp. 1–17, and J. Christensen, *Practicing Enlightenment: Hume and the Formation of a Literary Career*, Madison and London 1987.

[10] Very aptly for my purposes, Harold Bloom defines the canon as either 'the choice of books in our teaching institutions' – in its original meaning – or as 'the relation of an individual reader and writer to what has been preserved out of what has been written' (*The Western Canon: The Books and School of the Ages*, New York, San Diego and London 1994, pp. 15 and 17.)

not in terms of the influence of Locke's *Essay* on Hume's *Treatise*, but rather through the mediation of theoretically less charged books, such as eighteenth-century dictionaries, natural-philosophical treatises, or textbooks – or even novels – which have the advantage both of having been widely read and highly thought of then, and of being not very familiar to militant philosophers now. (And, when citing books still somewhat popular today, such as, for example, Bayle's *Dictionary* or the Port-Royal *Logic*, I shall sometimes use period editions and translations rather than the results of the most recent scholarly efforts.)

There is, of course, another kind of 'influence' which needs considering when one has to do with canonical authors. Hume is no longer the same after having been read by, say, Kant, Husserl, Ayer, and Kemp Smith, Laird, Passmore, Bennett, Livingston, Baier. The Hume read by me, by us, is partly constituted by them.

To me, Hume is of interest, for a start, as the author of a significant eighteenth-century book – a book as significant and even as entertaining as *Tristram Shandy* or *Tom Jones*, and as far from (or as close to) our present philosophical concerns as those are far from (or close to) the forms of our present moral and social life. But that the *Treatise* is interesting now is a fact. Why? It is tempting to suggest that 250 years are not enough to change human nature and its innate metaphysical instinct – to put it in the terms of the *Treatise*. On the other hand, it is also a fact that Hume's philosophical works (including the *Treatise*) have been read and commented on extensively and continuously during these 250 years, and have accordingly contributed to our own human nature and metaphysical instinct. It may well be that such differently oriented Hume readers can now so easily meet and, if necessary, so politely agree to disagree, because there is something deeper than their Hume interpretations that they share – the historically determined standards of, for example, historical fairness and analytic rigour, as well as the tradition of fair discussion and polite disagreement, to the establishment of which Hume's writings have so greatly contributed.

PART I

The two parts of Hume's system of space: the centrality of the self

CHAPTER 1

Reality and the coloured points: Sections 1–2 against infinite divisibility

So, Nat'ralists observe, a Flea
Hath smaller Fleas that on him prey;
And these have smaller still to bite 'em,
and so proceed ad infinitum.
(Jonathan Swift, *On Poetry*, 337)

But besides this exception, it may not be amiss to remark on this head, that the principle of the priority of impressions to ideas must be understood with another limitation, viz. that as our ideas are images of our impressions, so we can form secondary ideas, which are images of the primary; as appears from this very reasoning concerning them. This is not, properly speaking, an exception to the rule so much as an explanation of it. Ideas produce images of themselves in new ideas . . .

(*T*/6–7)

. . . and so *ad infinitum.*
And the great fleas themselves, in turn,
have greater fleas to go on;
While these again have greater still,
and greater still, and so on.
(A. De Morgan, *Budget of Paradoxes*)

INTRODUCTION

This chapter is devoted to interpreting the following lines:

When you tell me of the thousandth and ten thousandth part of a grain of sand, I have a distinct idea of these numbers and of their different proportions; but the images, which I form in my mind to represent the things themselves, are nothing different from each other, nor inferior to that image, by which I represent the grain of sand itself, which is suppos'd so vastly to exceed them.

> What consists of parts is distinguishable into them, and what is distinguishable is separable. But whatever we may imagine of the thing, the idea of a grain of sand is not distinguishable, nor separable into twenty, much less into a thousand, ten thousand, or an infinite number of different ideas. (*T*/27)

In this passage, taken from Part 2, section 1, 'Of the infinite divisibility of our ideas of space and time', there are three different elements, which refer to, or suggest, three different levels of discourse: the 'distinct ideas' of the numbers and proportions; the 'inseparable', that is, indivisible, image representing both the grain of sand and its parts; and 'the things themselves', the grain of sand and its parts. The image representing the grain of sand does not differ from that representing its thousandth or ten thousandth part; hence the idea of a grain of sand, meant as a mental image, is not divisible into parts. The ideas of the numbers and of the proportions between them, however, are said to be 'distinct'. So, the idea of a thousandth or ten thousandth part of a grain of sand cannot be imagined, but it can be conceived in terms of a relation. The mention of the 'things', finally, plays an important strategic role in the organisation of the discourse as a whole. Apparently, it only serves to remind the reader that whatever one 'imagines' about reality is irrelevant to the problem here under discussion; but, in fact, it emphasises the nature and the terms of the problem itself: the positive existence of perceptual minima – in this case 'ideas', or, more accurately, minima of imagination.

Even more importantly, the mention of 'the thing' implies that our image of a grain of sand is, in one specific sense, inadequate: it alludes to the fact that we know already, no matter how, that a grain of sand does consist of parts – that it is not indivisible. The image of the grain of sand is meant to be pictorial, modelled on the immediate data of visual perception. And visual perception constitutes the fourth, latent element and level of discourse in the quotation. Sight, as the present case illustrates clearly, when taken literally is misleading, because it sometimes finds the indivisible units in what we know to be the wrong place. At the same time, since it does indeed give us impressions which cannot be divided, it is the model of, and affords the language for, talking about elementary impressions and ideas in general.

So, there are four problems in the quotation: how Hume establishes the relationship between ideas and mental images; how he establishes the relationship between mental and visual images; to what extent and in what way he allows the possibility of conceiving ideas to which no image corresponds; and how far he goes in the supposition of external reality.

These four problems are crucial to the theory of the understanding in the *Treatise*, and are frequently and extensively discussed in the literature. In examining them in connection with the treatment of the ideas of space and time, Kemp Smith concentrates on Hume's answer to Bayle's scepticism and on the extensionlessness of points, which he relates to the ambiguous notion of the idea of space as 'manner of appearance'.[1] Laird emphasises Hume's debt to Berkeley for the arguments on the *minimum sensibile*, and to Bayle for the terms and expressions in which he gives his own version of them. Laird, Maund, and Flew maintain that the crucial feature of Hume's discussion is the identification of conceiving with imagining and forming a mental image, so that an idea of infinity would not be conceivable because it would have to be an infinite idea.[2] This interpretation accounts for the most difficult aspects of Hume's discussion of the 'doctrine of infinite divisibility' (as well as, for instance, of his treatment of general ideas) in terms of a fundamental inadequacy of his implicit metaphysical assumptions. These interpreters (with some reservations in the case of Flew) consider the reference to external reality and the admission of the possibility of being deceived by the senses as signs of a looseness in Hume's commitment to phenomenalism, and as a major inconsistency within his thought.[3]

Realism is one of the major issues in recent Hume scholarship. The discussion of Hume's notions of conceiving and supposing and of his position concerning the existence of external objects and our belief in it, in particular, is related to my present subject in several ways. Some interpreters have focussed on Hume's use of physiological notions in tackling cognitive problems. Anderson finds a metaphysically realist position in the way Hume accounts for perceptions as representations. On this basis, he maintains that Hume's discussion of sense-perception, with particular reference to sight and to extension, shows a clear physiological materialism: the ideas of extended bodies, for instance,

[1] N. Kemp Smith, *The Philosophy of David Hume*, London 1941, pp. 276ff., for the extensionless minima; pp. 284ff., on Hume's answering Bayle; and pp. 273ff., for the discussion of space as a manner of appearance. Cf. A.-L. Leroy, *David Hume*, Paris 1953, pp. 155–6. The derivation of the idea of space from 'manner of appearance' is the subject of chapter 2.

[2] C. Maund, *Hume's Theory of Knowledge*, London 1937, pp. 169, 199; J. Laird, *Hume's Philosophy of Human Nature*, London 1932, pp. 67–8; A. Flew, 'Infinite divisibility in Hume's *Treatise*', in D. W. Livingston and J. T. King (eds.), *Hume: A Re-evaluation*, New York 1976, pp. 257–69.

[3] Maund, *Knowledge*, pp. 200ff.; Laird, *Hume's Philosophy*, p. 68; Flew, 'Infinite divisibility', p. 261, writes that Hume 'provides no answer' to the question: 'how is it supposed to make sense to compare the relative sizes of ideas and impressions, on the one hand, with those of physical objects, on the other?'

would themselves be extended.[4] Wright emphasises the role of psychophysiological processes as the concrete basis for the operations of imagination. The reference to anatomical theories of the time (mainly Malebranche's version of the Cartesian position, widely received in Britain) is, he suggests, the key to Hume's philosophy as a whole, that is, of its balance between a sceptical theory of knowledge and a realistic metaphysics.[5] Arguing against Anderson's interpretation, Yolton first examines the role of visual language in seventeenth- and eighteenth-century studies of cognitive problems, then goes on to trace visual imagery in Hume's theory of perceptual acquaintance and concept formation, and in his whole account of mental activities.[6]

Certain aspects of the most recent discussions of realism are also connected with my present problem through the notion of 'relative ideas'. Flage, author of the most detailed study of relative ideas in Hume, agrees that the main sense of 'idea' in Hume's writings is 'image'. But he also points out passages in the *Treatise* where Hume refers to ideas or conceptions without any correspondent image: they are, according to Flage, 'relative ideas' of unperceived objects, and he describes them as a cognitive analogue of Russell's definite descriptions. The ideas of the thousandth and ten thousandth part of a grain of sand are relative ideas, as are the idea of the missing shade of blue, the idea of a substance, and even the idea of God.[7] Other interpreters, Galen Strawson and Wilson, for example, link the notion of relative ideas directly to the suppositions about the existence of external objects different from our perceptions, that is, to a realistic presupposition that they emphasise to the point of making Hume into a metaphysical realist. This realist philosopher is the so-called 'new Hume'.[8]

[4] R. F. Anderson, 'The location, extension, shape and size of Hume's perceptions', in Livingston and King (eds.), *Re-evaluation*, pp. 153–71.

[5] J. Wright, *The Sceptical Realism of David Hume*, Manchester 1983, esp. pp. 68ff. and 204ff.

[6] J. W. Yolton, *Perceptual Acquaintance from Descartes to Reid*, Oxford 1984, and 'Hume's ideas', *Hume Studies* 6 (1), 1980, pp. 1–25, followed by Anderson's reply in pp. 26–31. See also Yolton's review of E. Craig's *The Mind of God and the Works of Man*, Oxford 1987 in M. Stewart (ed.), *Studies in the Philosophy of the Scottish Enlightenment*, Oxford 1990, pp. 306–11.

[7] D. Flage, *Hume's Theory of Mind*, London and New York 1990, chapter 3, pp. 39–60, especially pp. 47ff. ; see also his 'Hume's relative ideas', *Hume Studies* 7, 1981, pp. 55–73. See also M. Thomas, 'Relative ideas rejected', *Hume Studies* 8, 1982, pp. 149–57, followed by Flage's reply, 'Relative ideas revisited', ibid., pp. 158–71.

[8] G. Strawson, *The Secret Connexion*, Oxford 1989, pp. 50ff. (cf. 121ff., 127ff.); W. Wilson, 'Is Hume a sceptic with regard to the senses?', *Journal of the History of Philosophy* 27 (1), 1989, pp. 49–73. See also Craig, *Mind of God*, pp. 123ff. Against this interpretative line, see the careful (and to me convincing) reading proposed by D. Livingston, 'A Sellarsian Hume?', *Journal of the History of Philosophy* 29 (2), 1991, pp. 281–90; see also his *Hume's Philosophy of Common Life*, Chicago 1984, esp. pp. 155ff. For an account of the recent discussion of Hume's realism see K. P. Winkler, 'The new Hume', *The Philosophical Review* 100 (4), 1991, pp. 541–79.

In my opinion, the general theme of the first section of Part 2 of the *Treatise* is the adequacy of our perception, which is a pervasive basic assumption of his philosophising. In the frame of reference constituted by the *Treatise*, any serious doubt about this adequacy is unreasonable, perhaps even inconceivable; and yet, in sections 1–2, Hume launches into a defence of it. The result is a highly paradoxical piece of writing. One of its paradoxes appears in connection with the argument of the grain of sand: from the way it is presented, it is evident that the case of the grain of sand is meant to be read as an argument *against* the doctrine of infinite divisibility; but why and how this is the case is far from obvious from the way it actually goes.

The aim of this chapter is to solve the paradoxes involved in the case of the grain of sand – or, to put it in more general terms, to identify both Hume's argument against the doctrine of infinite divisibility and the assumptions on which it rests. The complexities of section 1 of the *Treatise*, together with its alien frame of reference, are such, however, that I shall first have to take on a series of more local problems. These will occupy the following pages, in which I consider, in turn: Hume's far from simple way of articulating the relation between 'idea' and 'image', that is, between 'conceiving' and 'imagining'; his use of a quasi-Cartesian philosophical idiom to introduce the discussion of the doctrine of infinite divisibility; his description of the units of sight as unextended coloured points, and his account of their role in the formation of ideas; and, finally, his quick dismissal of the distinction between aliquot and proportional parts as a solid foundation for a doctrine of infinite divisibility. Only then shall I be in a position to tackle the central issue by discussing the nature of perceptual minima, by setting out and assessing some rival articulations of Hume's argument, and by defending one of them.

IMAGES AND IDEAS

I have mentioned the standard criticism of Hume's approach: that he equates conceiving with imagining and imagining with forming a mental image, or – which is basically the same, but more directly pertinent to my present discussion – that according to Hume an idea of infinity must be an infinite idea. It may be useful to clarify this point straightaway: in my opinion, Hume does not identify conceiving and imagining at all. What he does is, I think, much more complex.

Let me stick to the case of the grain of sand. It is evident that the emphasis on visual terminology does not mean that only what can be expressed directly in visual terms is legitimate knowledge. This is shown by, for instance, the reference to the 'distinct idea of these numbers and of their different proportions'. This reference is consistent with some other passages of the *Treatise*. In particular, concerning small and big numbers Hume writes:

> when we mention any great number, such as a thousand, the mind has generally no adequate idea of it, but only a power of producing such an idea, by its adequate idea of the decimals, under which the number is comprehended. This imperfection, however, in our ideas, is never felt in our reasonings; which seems to be an instance parallel to the present one of universal ideas. (*T*/22–3)

As others have remarked,[9] the lack of adequate ideas of great numbers, such as a thousand, and the inability to form any image specifically representing a thousandth part of a grain of sand – and not of both the grain of sand itself and, say, its ten thousandth part as well – are certainly related: one cannot help noticing that even the number chosen as an illustration of the two cases is the same. Hume calls this an 'imperfection in our ideas'. But he also says that it is of no consequence, and, in fact, assimilates it to the operation which produces 'universal ideas'. Our limited minds can only conceive particulars, and yet the 'abstract ideas' and 'general terms' available to us are perfectly efficient reasoning tools: they are, in fact, ideas which 'are particular in their nature', but become 'general in their representation' 'by being annex'd to a general term' – that is, to a term which is in turn made general by a 'customary conjunction' with many particular ideas and 'readily recalls them in the imagination' (*T*/22). Similarly, our reasonings about large numbers are not affected by our inability to conceive them adequately. The relation of quantity and number, one of the seven 'philosophical relations', is enough to let us know what we are talking about when we mention a thousand.

The similarity between the case of the great numbers and that of the grain of sand is evident; however, the passage on the grain of sand differs from that on great numbers in two important respects. First, the conception of the numbers and proportions involved in the case of the grain of sand is dependent on the mechanism of conception of large numbers, in the sense that, if we can conceive large numbers, it is clear that there are no particular problems in conceiving orders of

[9] See, for example, Flage, *Hume's Theory*, p. 49.

magnitude as well. Second, in the case of the grain of sand, the relationship between idea and image is not as easily established as is that between adequate idea and power of producing it in the case of a great number. There is an evident difference of emphasis between the two cases: while talking about the grain of sand, Hume does not say that the distinct ideas of the numbers and their different proportions, acquired and justified through the relation of quantity and number, are all we need to make sense of the situation; and it is precisely the contrast between the distinct ideas of the numbers and the impossibility of forming specific images of the objects discussed which is the most important point.

In order to make sense of the relationship between image and idea, let us follow Yolton's suggestion and start wondering what an 'image' is in the language of the *Treatise*.[10] The first point to be noticed is that 'image' obviously relates to 'imagination', the most obscure and ambiguous notion of Hume's theory of the understanding. It is frequently associated with 'fancy': 'the loose, floating images of the fancy' (*T*/116), and the 'images . . . presented to our fancy' (*T*/121) are quite common expressions in the *Treatise*, together with such similar ones as 'images . . . always wandering in the mind' (*T*/119) or even 'the loose conceptions of the imagination' (*T*/106). Following up this association reveals that 'image' is, like 'imagination', a central notion in Hume's discussions, and also that it is, again like 'imagination', far from clearly and unambiguously determined. For instance, concerning the effect of other people's testimony on belief, one finds that 'other effects only point out their causes in an oblique manner; but the testimony of men does it directly, and is to be consider'd as an image as well as an effect' (*T*/113). Also, if on the one hand there are passages where 'image' is used for, or seems to be specifically related to 'idea', as in the discussion of the grain of sand, on the other there are also such passages as: 'some ideas, which we form in the fancy; and images, which appear to the senses' (*T*/28). And, in this passage, not only is 'image' associated with the senses, so that the term is used synonymously with 'impression', but also 'idea' is associated with 'fancy', which, from what I was saying above, one might expect to find together with 'image'.[11] Apparently 'image', 'idea', 'impression', 'fancy', 'imagination', and so on, are not a

[10] See Yolton, 'Hume's ideas', pp. 2ff.; and also R. J. Butler, 'Hume's impressions', in G. Vesey (ed.), *Impressions of Empiricism*, London 1976, pp. 122–36, esp. pp. 123–4.

[11] 'Image' as synonym of 'impression' is also present in other passages of the *Treatise*; for instance, in *T*/205 we find: 'the very image, which is present to the senses'.

set of technical terms, but a group of fuzzy notions intertwined in complex and puzzling ways.

With reference to 'idea' in particular, there is another use of 'image' that needs considering and that may be clarified by what I have just said. Typically ideas are said to be 'images' of impressions:

> by ideas I mean the faint images of these [impressions, i.e. all our sensations, passions, and emotions] in thinking and reasoning; (*T*/1)

> as our ideas are images of our impressions, so we can form secondary ideas, which are images of the primary; as appears from this very reasoning concerning them. (*T*/6)

Another term used to describe the relation of ideas to their correspondent impressions is 'copies' (*T*/3, cf. *T*/72, 96). A literal reading of these passages would yield a theory of concept formation entailing a game of mirrors: impressions of the senses and passions being reflected in primary ideas, primary ideas being reflected, in turn, in secondary ideas. The key to such a theory would be, obviously enough, the relation of resemblance, which would be responsible for the whole structure. This is the case according to Anderson: in particular ideas related to impressions of sight, that is, ideas of extended bodies are, he claims, to be read as themselves extended like the impressions they are supposed to resemble.[12]

Now, there are several problems with this reading. At the moment, I shall consider only one of them. The relation of resemblance, as it is presented by Hume in the *Treatise*, neither implies nor justifies this kind of complete mirroring. For resemblance is compatible with a degree of difference between related perceptions: for example, Hume maintains that in comparing our perceptions we may well find them resembling each other in only one respect (*T*/25), and even says that resemblance may be found between simple ideas (*T*/637). Resemblance establishes a relation by singling out specific aspects of different perceptions. So, when Hume talks of the resemblance of ideas to their correspondent impressions, there is no implication that such a resemblance must be a mirroring. Nor is there any implication of this sort even when Hume says that the only difference between an idea and its correspondent impression is the 'degree of force and vivacity'. On the other hand, ideas are not only 'images' or 'copies' of impressions. They are also defined as 'exact representations' (*T*/3) or 'copies and representations'

[12] Anderson, 'Location', pp. 163ff., esp. p. 165.

of impressions (*T*/19); they 'are deriv'd from, and represent impressions' (*T*/161; see also *T*/7); an idea is 'borrow'd from, and represents some impression' (*T*/34). Also, while stating that the distinction between impressions and ideas is familiar to everybody, Hume describes it as 'the difference between feeling and thinking' (*T*/2). In short, one is faced again with a constellation of terms whose mutual relations are far from straightforward. It seems reasonable to conclude, in general, that talking of ideas as 'images' of impressions expresses a link rather than literally describing a correspondence, and that there is no reason to think that ideas literally mirror impressions. This reading of Hume's use of 'image' – among other terms – to describe the relation between impressions and ideas is in agreement not only with what is suggested by the other uses of 'image', but also with the use of other terms – for instance, 'impression' itself. 'Impression' is consciously taken by Hume in a metaphorical sense, not

> to express the manner, in which our lively perceptions are produced in the soul, but merely the perceptions themselves; for which there is no particular name either in the *English* or any other language, that I know of. (*T*/2 fn.)[13]

Let me return to the grain of sand. Yolton rightly remarks that 'the use of "image" in this passage is not made clear'[14] – and I have shown that this particular problem is anything but peculiar to the case of the grain of sand. Here the use of 'image' is more evidently related to the literal meaning of the term than in any of the passages considered above. It is not difficult to find a good reason for this: Hume is discussing a problem related to the idea of extension. It is not surprising that the emphasis is shifted in the direction of sight, since it is (mainly) from impressions of sight that the idea of extension is derived: 'the idea of extension is nothing but a copy of these colour's points, and of the manner of their appearance' (*T*/34).[15] However, some pages later one finds:

[13] Cf. Locke, *Essay*, II.i. sect.23: 'Sensation is such an *impression* or motion, made in some part of the body, as produces some perception in the understanding.' On Hume's use of 'impression' and on the meaning of the term in scholastic theories of perception, in the Cartesian and Malebranchean theory of animal spirits, in Samuel Clarke, and in Locke's *Essay*, see Laird, *Hume's Philosophy*, pp. 26ff. 'Impression' also belongs to other domains – for example, to the language of a printer's shop. On the basis of Locke, in his *Dictionary* Johnson gives as first definition 'The act of pressing one body upon another' (and then, in order: 'mark made by pressure; stamp', 'image fixed in the mind', 'operation; influence', 'edition; number printed at once; one course of printing', 'effect of an attack').

[14] Yolton, 'Hume's Ideas', p. 7.

[15] Cf. F. Zabeeh, *Hume Precursor of Modern Empiricism*, The Hague 1960, p. 72.

> That compound impression, which represents extension, consists of several lesser impressions, that are indivisible to the eye or feeling, and may be call'd impressions of atoms or corpuscles endow'd with colour and solidity. (*T*/38)

The reference to the units of touch in this passage is revealing. In discussing space, Hume talks mainly about the units of sight and their property of being coloured, for the obvious reason that they can be more easily singled out; but what he has in mind concerning his perceptual units is not only colour and not only sight. Second, it is worth noticing that, if one concentrates on the visual impressions of the points and the way they are described in this passage, one finds, implicit, the distinction between the idea of a coloured point and the idea of its particular colour. These two facts indicate a tendency toward abstraction, in agreement with the use of visual language in a metaphorically extended sense.

As I shall show, Hume's language is borrowed from geometrical optics and the anatomy of the eye, and is used to describe perceptual minima which are in turn the model for the elementary contents of the mind. What are the features of visual images that make them so suitable to provide the language for the treatment of mental contents, even of an entirely different kind? How does the use of a visual metaphor affect the discourse on such mental contents? Or, in other words: what is the relationship between the 'moments of the mind' and the indivisibles of space?

But another problem is now more urgent. Why is it that visual metaphors in Hume's text suggest such a strong commitment on his part to the identification of idea and mental (pictorial) image? After all, visual language is very commonly employed to describe mental operations – think of 'clarity' and 'distinctness'. It is my belief that Hume's text is so organised as to cause the reader to bear constantly in mind the literal meaning of his visual metaphors – so that, for instance, while reading Hume's page one tends somehow to refer the notion of the clarity of an idea to such notions (or images?) as the colour of a visual point. This is the effect of a complex of rhetorical moves, which now needs examining.

MANNER AND MATTER IN SECTION I

The immediate aim of Hume's discussion of perceptual points in section 1 is to demonstrate that the 'doctrine of infinite divisibility' is wrong, or, as he puts it in section 2, that 'all the pretended demonstrations for the infinite divisibility of extension are equally sophistical' (*T*/33).

Hume's tone in introducing his discussion of the 'doctrine of infinite divisibility' and in arranging this first section on the subject is simple and reassuring: 'Whatever has the air of a paradox, and is contrary to the first and most unprejudic'd notions of mankind is greedily embrac'd by philosophers', so that they 'furnish . . . plenty of strange and unaccountable opinions' – 'of which I cannot give a more evident instance than the doctrine of infinite divisibility' – and their disciples 'readily believe them'. In contrast with this procedure, Hume claims that his approach is based on a fact which is 'universally allow'd . . . and tho' it were not allow'd, 'twou'd be sufficiently evident from the plainest observation and experience' (*T*/26). This matter of common knowledge is the limited capacity of the mind. And it is with this appeal to what is (apparently) evident, plain, and obvious that Hume opens one of the most complicated and difficult discussions of the whole *Treatise*.

Even the introduction of the limited capacity of the mind presupposes a sophisticated theoretical frame of reference: ''tis universally allow'd, that the capacity of the mind is limited, and can never attain a full and adequate conception of infinity' (*T*/26). Laird remarks on the similarity between this statement and the Port Royal axiom that 'Il est de la nature d'un esprit fini de ne pouvoir comprendre l'infini',[16] and emphasises that this statement of the limited capacity of the mind is not a matter of common sense, as Hume appears to claim, but a metaphysical principle within the Cartesian tradition, and in addition one that was anything but 'universally allow'd'. For instance, he reminds the reader that 'Mens nostra, eo quod finita sit, nihil certo scire potest de infinito' was among the Cartesian propositions condemned by the Jesuits. On the other hand, the emphasis on the process of 'attaining a conception' in Hume's text refers to another, in fact entirely different,

[16] A. Arnauld and P. Nicole, *La Logique ou l'Art de Penser*, ed. by P. Clair and F. Girbal, Paris 1965, P. IV, ch. 7, ax. 9, p. 322 [in the translation of J. Dickoff and P. James, *The Art of Thinking*, Indianapolis 1964, p. 324: 'The nature of a finite mind is such that it cannot grasp the infinite'; in a period translation: 'It is the Nature of a final [*sic*] Spirit not to apprehend the Infinite' (*Logic; or, the Art of Thinking*, London 1685, p. 204)]; cf. Descartes, *Principia Philosophiae*, I, 26, ed. by Adam and Tannery, vol. 8, Paris 1964, p. 14: 'Ita nullis unquam fatigabimur disputationibus de infinito. Nam sanè, cùm simus finiti, absurdum esset nos aliquid de ipso determinare, atque sic illud quasi finire ac comprehendere conari' [in the translation by E. S. Haldane and G. R. T. Ross, *The Philosophical Works of Descartes*, vol. 1, Cambridge 1911, p. 229: 'We will thus never hamper ourselves with disputes about the infinite, since it would be absurd that we who are finite should undertake to decide anything regarding it, and by this means in trying to comprehend it, so to speak regard it as finite'].

kind of interest and approach. As Newman points out, for instance, Hume's formulation of the inconceivability of infinity closely recalls Locke's way of presenting infinity as a 'growing and fugitive idea, still in a boundless Progression, that can stop no where'. So, Hume is using a Cartesian principle, and is rephrasing it in such a way as to divorce it entirely from any implication about certainty and immediate intuition and to shift it well within the inquiry into the origin of ideas. The result is a blend of rationalistic metaphysical principles with empiricist attitudes which constitutes a highly technical language, and it is this very language Hume speaks throughout sections 1 and 2. It is a sort of dissonant Cartesianism, so that, to take another salient illustration, the principle that 'whatever the mind clearly conceives includes the idea of a possible existence' is translated, through the clause 'or in other words', into 'nothing we imagine is absolutely impossible'.[17] But this move needs explaining at a different level; for, as I am going to show, its consequences can hardly be overestimated.

Hume's next step towards the refutation of the doctrine of infinite divisibility is to introduce a second allegedly obvious fact:

> 'tis also obvious, that whatever is capable of being divided *in infinitum*, must consist of an infinite number of parts, and that 'tis impossible to set any bounds to the number of parts, without setting bounds at the same time to the division. (*T*/26–7)

Much needs to be considered before one can deal with the many problems involved in this statement. At the moment I shall discuss only one point: commentators have claimed that this second 'obvious' fact Hume is relying on, though certainly quite common in the period, is no more obvious than the first one. Flew, for instance, maintains that it is 'without qualification false', because it

> implies that [infinitely divisible] finite things are constituted of infinite collections of other finite things, which in turn are constituted of infinite collections of infinitely smaller finite things, and so on, *in infinitum*;

while to say that something is infinitely divisible

[17] See Laird, *Hume's Philosophy*, pp. 67–8, for a discussion of Hume's French sources. C. W. Hendel, *Studies in the Philosophy of David Hume*, 2nd edition, Indianapolis 1963, p. 456, commenting on section 2, says that Hume 'writes in true Cartesian style'. R. Newman, 'Hume on space and geometry', *Hume Studies* 7, 1981, pp. 1–31, esp. p. 4, suggests the comparison between Hume's formulation of the limited capacity of the mind and Locke's definition of infinity (which appears in *Essay*, II.xvii. sect.12).

is not to say that it can be divided into an infinite number of parts. It is rather to say that it can divided, and sub-divided, and sub-sub-divided as often as anyone wishes: infinitely, without limit.

Flew concludes that

the contradictions and absurdities, whether real or only apparent, which make the doctrine of infinite divisibility scandalous to Hume spring precisely from this proposition [. . .it] is the very misconception which generates the paradoxes he wishes to remove.[18]

Now, instead of, or at least before, deciding whether Hume's statement of the infinity of an infinitely divisible extension is right or wrong and why, one may usefully ask: what is its function (and hence its concrete meaning) in Hume's argument? Or, to put it differently: how does the case of the grain of sand refute the doctrine of infinite divisibility?

As I have said, giving a full answer to the latter question may be considered the general aim of the present chapter. A first step towards answering this question is again to wonder about words – this time, in particular, about the expression 'doctrine of infinite divisibility'. As I have already observed, the main theme of section 1 is the adequacy of our perception. What the doctrine of infinite divisibility represents here, in the economy of Hume's argument, is the shift from the limited capacity of the mind to its inadequacy; in addition, this inadequacy is allegedly proved through demonstrative arguments. Both this way of seeing the doctrine of infinite divisibility and the notions of division and of parts implied in Hume's statement are very common; indeed, they are almost commonplace amongst mathematicians and metaphysicians of the period.[19] We find them, for instance, both in Barrow and in Malezieu. Thus Barrow writes:

[18] Flew, 'Infinite divisibility', p. 260.

[19] These notions are also shared by other major philosophers of the period – in particular by Berkeley (see below, fns. 25 and 41–2, for some examples). The comparison of Hume's views on these matters with Berkeley's is standard in the literature: see, for example, Laird, *Hume's Philosophy*, pp. 68ff. For a more recent, full examination of the influence on Hume of Berkeley's treatment of space, infinity, and perception see M. Ayers, 'Berkeley and Hume: a question of influence', in R. Rorty, J. B. Schneewind, and Q. Skinner (eds.), *Philosophy in History: Essays on the Historiography of Philosophy*, Cambridge 1984, pp. 303–27 (with further references to the 1960s debate about whether Hume ever read Berkeley). See also D. Raynor, '"*Minima sensibilia*" in Berkeley and Hume', *Dialogue* 19, 1980, pp. 196–9; R. Fogelin, *Philosophical Interpretations*, Oxford 1992, chapter 3, 'Hume and Berkeley on the proofs of infinite divisibility', pp. 45–69; and J. Franklin, 'Achievements and fallacies in Hume's account of infinite divisibility', *Hume Studies* 20 (1), 1994, pp. 85–101. Another obvious reference is Bayle (whose influence is emphasised in

I deny not but it is difficult to be understood, how every single Part can be divided so as all not to be actually reduced by the Division to Indivisibles, or to Nothing or what is next to Nothing: Nor yet do I think, by reason of the Imperfection of the Mind of Man and the Smallness of our Capacities, that therefore the Truth is to be deserted, when proved by so many evident Tokens, and supported by so many strong Arguments . . . we clearly perceive it [this indefinite Division] to follow of necessity from the Nature of Matter, a Thing most manifestly known to us, and we do also perceive it to be of that kind of Things, which cannot be comprehended by our Minds, as being finite. It follows therefore that every Quantity is compounded of Parts that are compounded, and may be divided into Parts that are again divisible.[20]

Similarly (but with a direct invocation of God), Malezieu says:

comme l'esprit humaine est borné, et que le Createur de nos ames, ne leur a pas donné des lumieres infinies, c'est à nous de nous souvenir de notre condition . . . On a donc grand tort de vouloir attaquer la Geometrie des infiniment Petits, et celle des Indivisibles, parce qu'il y a de certaines choses qu'on ne comprend pas dans la nature de l'infinie, qui en effet doit etre incomprehensible.[21]

Malezieu's text roughly repeats Arnauld's words:

notre esprit étant fini, il se perd et s'éblouit dans l'infinité, et demeure accablé sous la multitude des pensées contraires qu'elle fournit . . . toute la vigueur de l'esprit des hommes est contrainte de soccomber au plus petit atome de la

Kemp Smith, *Philosophy*, chapter 14, and Laird, *Hume's Philosophy*, pp. 78ff.): see below, in the next section, 'Units of sight'.

On Hume's acquaintance with mathematical texts and doctrines see Kemp Smith, *Philosophy*, pp. 339ff. for Malezieu, and pp. 343ff. for Barrow; and M. Frasca-Spada, 'Some aspects of Hume's conception of space', *Studies in History and Philosophy of Science* 21 (3), 1990, pp. 371–411. (For a later period see also L. Gossmann, 'Two unpublished essays on mathematics in the Hume papers', *Journal of the History of Ideas* 21, 1960, pp. 442–9.) M. A. Stewart, *Introduction*, in Stewart (ed.), *Studies in the Philosophy of the Scottish Enlightenment*, Oxford 1990, pp. 1–9, esp. pp. 8–9, and M. Barfoot, 'Hume and the culture of science in the early eighteenth century', in *ibid.*, pp. 151–90, esp. p. 190, describe a very early Humean manuscript, 'A treatise of fluxions by Mr George Campbell Professor of Mathematics in Edinburgh', dated 1726, containing a set of lecture notes on the principles of the Newtonian calculus (this MS was sold as lot 473 at Sotheby's sale of *English Literature and History*, London 21–2 July 1988, see the saleroom catalogue, p. 241).

20 I. Barrow, *The Usefulness of Mathematical Learning Explained and Demonstrated. Being Mathematical Lectures Read in the Public Schools at the University of Cambridge* by Isaac Barrow, D.D. Professor of Mathematics, and Master of the Trinity College, etc. Translated by the Revd Mr. John Kirkby, of Egremont in Cumberland, London 1734 (facsimile edition London 1970), p. 162.

21 N. Malezieu, *Elemens de Geometrie de Monseigneur le Duc de Bourgogne*, Troisième edition, Revûë, corrigée & augmentée d'un Traité des Logarithmes, Par M. De Malezieu. Avec l'Introduction a l'Application de l'algebre a la geometrie, Paris 1734, pp. 30–1 ['as the human mind is limited, and as the Creator of our souls has not given them infinite light, we must be reminded of our condition . . . It is therefore quite wrong to wish to attack the Geometry of the Infinitely Small and that of Indivisibles, because there are certain things in the nature of infinity which are not understood and which must in effect be incomprehensible'].

matiere, et d'avouer qu'il voit clairement qu'il est infiniment divisible, sans pouvoir comprendre comment cela se peut faire . . .[22]

Arnauld's mention of the 'plus petit atome de la matiere' suggests that even Hume's set of examples – a grain of sand, an insect a thousand times less than a mite – is revealing.[23] In the works quoted above, a 'doctrine of infinite divisibility' is associated with the complete inability of the mind to make sense even of something as simple and humble as a grain of sand or dust, a poppy seed, a mite, the minute creatures revealed by microscopy; for the mystery of God's overwhelming infinity is hidden everywhere in the universe. In both Malebranche and Arnauld each of these minute entities is the proliferation of an infinity of different worlds. Malebranche writes:

On voit assez souvent avec des lunettes des animaux beaucoup plus petits qu'un grain de sable qui est presque invisible; on en a vu même de mille fois plus petits . . . Les petits animaux dont nous venons de parler, ont peut-être d'autres petits animaux qui les dévorent, et qui leur sont imperceptibles à cause de leur petitess effroyable, de même que ces autres nous sont imperceptibles. Ce qu'un ciron est à notre égard, ces animaux sont à un ciron; et peut-être qu'il y en a dans la nature de plus petits, et de plus petits à l'infini dans cette proportion si étrange d'un homme à un ciron.[24]

And Arnauld and Nicole:

22 *Art de Penser*, Part IV, chapter 1, pp. 295 and 299 [in the translation by Dickoff and James, pp. 297 and 301: 'our minds are finite; and blinded by the infinite, they are lost in it – forever overwhelmed by the multitude of conflicting thoughts the infinite furnishes . . . the strongest mind must yield to the smallest atom of matter and acknowledge that the mind clearly sees matter to be infinitely divisible without understanding such divisibility'; in the translation London 1685, pp. 166 and 170: 'For our understanding being finite, looses it self in the Labyrinth of Infinity, and lies overwhelm'd under the multitude of thoughts, contradicting one another . . . the force and vigor of human wit is forc'd to succomb to the least Atom of matter, and to confess that it clearly sees, that it is divisible into Infinity, not apprehending how it can be done'].

23 The number of the appearances of the mite, the grain of sand etc. to similar purposes in the literature and philosophy of the seventeenth and eighteenth centuries is striking. See M. H. Nicolson, 'The microscope and the English imagination', in her *Science and Imagination*, New York 1956, pp. 155ff. (Nicolson quotes from Fontenelle in pp. 184–5; Algarotti in p. 189; Pascal, in p. 210; Pope in p. 219; W. Derham, who repeats Malebranche almost word for word, in p. 223; and so on).

24 *Recherche de la Vérité*, B. 1, ch. 6, section 1 [In the English transl. by T. M. Lennon and P. J. Olscamp, *The Search after Truth*, Columbus 1980, pp. 25–6: 'With magnifying glasses, we can easily see animals much smaller than an almost invisible grain of sand; we have seen some even a thousand times smaller . . . For the tiny animals of which we have just spoken, there are perhaps other animals that prey upon them and that, on account of their awesome smallness, are to them as imperceptible as they themselves are to us. What a mite is compared to us, these animals are to a mite; and perhaps there are in nature things smaller and smaller to infinity, standing in that extreme proportion of man to mite'].

cette partie dont la petitesse nous est déja incomprehensible, contient encore un autre monde proportionel, et ainsi à l'infini, sans qu'on puisse trouver aucune qui n'ait autant de parties proportionelles que tout le monde, quelque étendue qu'on lui donne. (pp. 296–7)[25]

The way Hume puts it in the *Enquiry* sounds like a parody of Malebranche and Arnauld – a parody made grim by Hume's evident distaste:

No priestly *dogmas*, invented on purpose to tame and subdue the rebellious reason of mankind, ever shocked common sense more than the doctrine of the infinite divisibility of extension, with its consequences; as they are pompously displayed by all geometricians and metaphysicians, with a kind of triumph and exultation. A real quantity, infinitely less than any finite quantity, containing quantities infinitely less than itself, and so on *in infinitum*; this is an edifice so bold and prodigious, that it is too weighty for any pretended demonstration to support, because it shocks the clearest and most natural principles of human reason. (*E*/156)

What, in the terse style of the *Enquiry*, takes the form of a parody is, in the more abstract and complex language of the *Treatise*, the subject of an entangled dialectical game. The role of Hume's two 'universally allow'd' and 'obvious' facts in this game can now be clarified.

As a preliminary it may be useful to outline the whole argument presented in section 1. Philosophers love paradoxes, which make them seem, or at least feel, different from common people and which gratify the normal tendency of the human mind towards what is intricate and amazing. The doctrine of infinite divisibility is a typical illustration of this. However, the problem of the composition of extension can be very easily solved by reference to 'universally allow'd' and 'obvious' facts: first, the human mind is limited and cannot form an adequate conception of infinity; second, whatever is infinitely divisible contains an infinite number of parts, so that the division of any of our ideas would arrive very fast at an indivisible element. This is what actually happens:

[25] *Art de Penser*, P. IV, chapter 1, pp. 296–7 [in the transl. by Dickoff and James, p. 298: 'that part whose smallness is already incomprehensible to us contains still another proportional world, and so on to infinity – without our discovering any part, no matter how small, that does not have as many proportional parts as does the whole world'; in the translation London 1685, p. 167: 'this part which is so incomprehensible to us, contains another proportionable world, and so *ad infinitum*; there being still no part which does not comprehend as many proportional parts as the world, how large soever we make it']. The mite appears in Berkeley's *New Theory of Vision*, where, in section 80, we are told that the *minimum sensibile* is the same for men and mites (see also *Hylas*, Dial. 1). Laird suggests Pascal, *Pensées*, 72 as another possible source. The most obvious train of thought leads, of course, to Swift's flea.

our imagination does present such minima, so that, for instance, the image of a grain of sand is the same as that of, say, one thousandth part of it. Our senses also present minima: a spot of ink on paper, withdrawn to increasing distances from us, will before disappearing altogether, appear as an indivisible coloured point. Symmetrically, microscopes and telescopes applied to bodies distant or minute enough to be invisible make such bodies visible by spreading the rays of light flowing from them. We cannot form any adequate idea of what is too large for our perceptual apparatus, but we do have the most minute perceptions: so that, while it is evident that we are able to imagine the smallest atom of the animal spirits of an insect one thousand times less than a mite, it may be doubted whether we can ever conceive the whole insect if we start from the enumeration of its minute parts – however clearly they be conceived or imagined.

Hume's polemical target, one can now see, is precisely defined through two claims: that concerning the limited capacity of the mind and hence the inconceivability of infinity, and that concerning the infinite number of parts in an infinitely divisible object. And, if this is the polemical target, the rhetorical device he chooses when introducing this paradoxical subject in terms of what is 'universally allow'd' and 'evident from the plainest observation and experience' makes very good sense. For Hume's main concern is to maintain that of course the human mind is not really baffled by grains of sand, poppy seeds, mites and fleas, and other, even more minute, entities. While maintaining this, Hume also insinuates it in his tone and in the way he approaches the subject: nothing here is particularly worrying, the most puzzling problem is going to be solved in terms of the most elementary reasonableness by reference to the most obvious and evident facts of common experience, etc. Understanding such entities – this is the implication – is only complicated because of the confusing way they are usually discussed by philosophers, victims of their arrogance and, ironically, of the most common human passions. And the complexities of the philosophers' discussions are reflected, as it were, in Hume's writing, in a controlled tension with the programmatic simplicity of his own tone. This is what one finds when, for instance, Hume says that our perception is perfectly adequate to represent 'the smallest atom of the animal spirits of an insect a thousand times less than a mite', and that the only difficulty 'lies in enlarging our conceptions so much as to form a just notion of a mite, or even of an insect a thousand times less than a mite' (*T*/28). Here the mite is no longer the figure of God's wondrous infinity

– it is just an insect. And thus it is the difficulty in conceiving the insect which is the true outcome of the complexities of the philosophers' discussions, rather than the difficulty in conceiving its smallest parts. The puzzling aspect of the example is the reflection of, and at the same time the answer to, the unreasonableness of such discussions – nobody (not even Malebranche and Arnauld) would maintain that we cannot conceive a mite.

This is only a first step. I shall show later that the two elements crucial for an account of Hume's rejection of the 'doctrine of infinite divisibility' are:

1. His account of the units of sight and their role in the formation of ideas;
2. The way he introduces and employs the already detected realistic presupposition underlying his argument.

I shall start with the examination of some of the features Hume ascribes to the units of sight.

UNITS OF SIGHT

As I have said, the image of the grain of sand is treated by Hume in pictorial, visual terms. He introduces the units of sight in two complementary ways: by analysing the common experience of vision in terms of the elements composing it; and by constructing an isolated unit experimentally. The most interesting aspect of this part of Hume's treatment is the relation he establishes between natural philosophy and a discourse on human experience. It is precisely this relation that makes it possible to solve the problem of the composition of extension, through the notion of unextended coloured (and tangible) points.

> When an object augments or diminishes to the eye or imagination from a comparison with others, the image and idea of the object are still the same, and are equally extended in the *retina*, and in the brain or organ of perception. The eyes refract the rays of light, and the optic nerves convey the images to the brain in the very same manner, whether a great or small object has preceded; nor does even the imagination alter the dimensions of its object on account of a comparison with others. The question then is, how from the same impression and the same idea we can form such different judgments concerning the same object, and at one time admire its bulk, and at another despise its littleness. (*T*/372–3)

What Hume is describing here, in the language of natural philosophy, is visual experience in its functional position within the complex set of interactions – with judgements, other sensations, emotions – whose totality constitutes actual human experience. The passage is introductory to the section on the passions of malice and envy. It is part of a long illustration of a basic feature of human nature: 'objects appear greater or less by a comparison with others' (*T*/375), a principle which explains, among other things, a feature of our experience of vision as well as the emotional mechanism of malice and envy. Clearly, the language of natural philosophy is here exemplary for the discourse on human experience: as Hume puts it himself,

> Those, who are acquainted with the metaphysical part of optics, and know how we transfer the judgments and conclusions of the understanding to the senses, will easily conceive this whole operation. (*T*/374–5)[26]

So, the problem in the passage quoted above is the division of labour between eye and judgement – what is it that the eye really sees; that is, what is it that can be considered as really given in the sense-impression? This is a classical problem in the writings of geometrical optics and anatomy of the eye of the period.[27] Hume's answer, given extensively in two other pages of the *Treatise*, is quite classical as well. The only element of the experience of vision directly dependent on the eye is the set of coloured points on the retina:

> 'tis commonly allow'd by philosophers, that all bodies, which discover themselves to the eye, appear as if painted on a plain surface, and that their different degrees of remoteness from ourselves are discover'd more by reason than by the senses. When I hold up my hand before me, and spread my fingers, they are separated as perfectly by the blue colour of the firmament, as they cou'd be by any visible object, which I cou'd place betwixt them; (*T*/56)

and, with a more explicit reference to the act of the judgement:

> 'tis universally allow'd by the writers on opticks, that the eye at all times sees an equal number of physical points, and that a man on the top of a mountain has no larger an image presented to his senses, that when he is cooped up in the narrowest court or chamber. 'tis only by experience that he infers the greatness of the object from some peculiar qualities of the image; and this inference of

[26] I am grateful to Annette Baier for a stimulating comment on this page.

[27] On the treatment of the problems of vision in geometrical optics and anatomy in eighteenth-century Britain, see Yolton, *Perceptual Acquaintance*, especially chapter 7, pp. 124–46 (cf. his 'As in a looking glass: perceptual acquaintance in eighteenth-century Britain', *Journal for the History of the Ideas* 40, 1979, pp. 207–34); M. Baxandall, *Patterns of Intention*, New Haven and London 1985, chapter 3, especially pp. 81ff.; also N. Daniels, *Thomas Reid's Inquiry. The Geometry of Visibles and The Case for Realism*, New York 1974, pp. 30ff.

the judgment he confounds with sensation, as is common on other occasions. (*T*/112)

Vision is organised like a painting, flat shapes composed of points of different colours on the retina. The only change is in the distribution of the colours of the points; for the whole set of points is invariable. We always see the same number of points, determined by the structure of our eyes.[28]

There are here two main problems relevant to my present concern. First, what is a point? The points seen by the eye are said to be 'physical' (*T*/112). Similarly, in the *Enquiry*, Hume writes: 'we must allow that there are physical points' (*E*/124 fn., sect. 12, p. ii). However, in another part of the *Treatise* we find that 'The system of physical points . . . is too absurd to need a refutation' (*T*/40). Now, this passage of the *Enquiry* has been supposed to imply a clear shift from the position of the *Treatise*, precisely because of the reference it contains to physical points.[29] Obviously this need not be the case, given that the expression also occurs in the *Treatise* itself. What is Hume trying to say? Secondly, what are the 'peculiar qualities of the image' which guide the operation of judgement concerning the greatness of the object? Some clues to the first of these problems and a hint at the answer to the second may be found in the passages where Hume introduces the units of sight taken in isolation.

Put a spot of ink upon paper, fix your eye upon that spot, and retire to such a distance, that at last you lose sight of it; 'tis plain, that the moment before it vanish'd the image or impression was perfectly indivisible. (*T*/27)

Hume is presenting what purports to be a simple experiment, in fact not much more than a fact of common experience, which shows how to single out one of the units of sight. The spot of ink appears again some pages later, and Hume's description becomes more extensive and sensitive:[30]

Put a spot of ink upon paper, and retire to such a distance, that the spot becomes altogether invisible; you will find, that upon your return and nearer approach the spot first becomes visible by short intervals; and afterwards

28 Cf. below, chapter 2, 'A philosopher's eye'. In his stimulating and delightful 'Achievements and fallacies', p. 97, Franklin reports eighteenth-century and present-day calculations of the size of a visual indivisible, and hence of the number of visual indivisibles in an inch.

29 See, for instance, Kemp Smith, *Philosophy*, pp. 286–7, and B. M. Laing, *David Hume*, 2nd edition, New York 1968, p. 106; cf. Flew, 'Infinite divisibility', pp. 268–9.

30 C. D. Broad, 'Hume's theory of space', *Proceedings of the British Academy* 47, 1961, pp. 161–76, criticizes Hume heavily for the treatment of the spot of ink in *T*/27.

becomes always visible; and afterwards acquires only a new force in its colouring without augmenting its bulk; and afterwards, when it has increas'd to such a degree as to be really extended, 'tis still difficult for the imagination to break it into its component parts, because of the uneasiness it finds in the conception of such a minute object as a single point. This infirmity affects most of our reasonings on the present subject. (*T*/42)

What emerges here is that the particular kind of simplification of the matter obtained through the allusion to a common experience, reported in the language of experimental natural philosophy, provides the means for a more refined treatment of the units of sight. For instance, there is the reference to the change in the intensity of colour and the stability of the image, and to the uneasiness of the imagination in conceiving very minute objects. The description of the phenomenon in terms of human existence and its specific notions moves the exemplary discourse of natural philosophy one step forward and makes it more subtle. There are now taken into consideration even 'the peculiar qualities of the image', which are at a different conceptual level from the mere set of points apprehended by the eye.

Concerning the spot of ink, where one point is taken in isolation, the main feature of Hume's points appears more prominently than before: they are coloured. The spot of ink is defined and identified through its colour – it is colour. One may observe that referring to colour is a very good way to identify a mathematical point intuitively. On the other hand, it is clear that it is the space of human existence which is under discussion; and a secondary quality is particularly suitable to afford the notions and the language to talk about human experience. The reference to a secondary quality in order to define the perceptual points has a crucial consequence: extension is a feature that is not necessary to define Hume's perceptual points – colour is enough. In fact, the idea of an isolated point not only does not involve, but, in fact, positively excludes, any idea of extension at all:

let us take one of those simple indivisible ideas, of which the compound one of *extension* is form'd, and separating it from all others, let us form a judgment of its nature and qualities. 'tis plain it is not the idea of extension. For the idea of extension consists of parts; and this idea, according to the supposition, is perfectly simple and indivisible. Is it therefore nothing? That is absolutely impossible . . . That compound impression, which represents extension, consists of several lesser impressions, that are indivisible to the eye or feeling, and may be call'd impressions of atoms or corpuscles endow'd with colour and solidity. (*T*/38)

So, when, in the *Treatise* as in the *Enquiry*, Hume is talking of 'physical points', he is, in fact, talking of coloured and tangible, but unextended, points; the crucial feature of physical points, and that which makes them absurd according to the *Treatise*, is that they are supposed to be extended. In fact, in the passage of the *Treatise* where he states the absurdity of the system of physical points, Hume explains: 'A real extension, such as a physical point is suppos'd to be, can never exist without parts, different from each other' (*T*/40); and similarly, in the footnote in the *Enquiry* quoted above, Hume is also very clear in explaining that the physical points he is allowing are

> parts of extension, which cannot be divided or lessened, either by the eye or imagination. These images, then, which are present to the fancy or senses, are absolutely indivisible, and consequently must be allowed by mathematicians to be infinitely less than any real part of extension. (*E*/124 fn.)

Thus, coloured and tangible perceptual points are to be conceived, in the *Enquiry* as well as in the *Treatise*, as a medium between (extended) physical points and mathematical points, both of them absurd and impossible. In the background lurk Arnauld and Bayle. In the *Art de Penser* we find:

> il n'y a rien de plus claire que cette raison, que deux neants d'étendue ne peuvent former une étendue, et que toute étendue a des parties; or en prenant deux de ces parties qu'on suppose indivisibles, je demande si elles on de l'étendue, ou si elles n'en ont point: si elles en ont, elles sont donc divisibles, et elles ont plusieurs parties; si elles n'en ont point, ce sont donc des neants d'étendue; et ainsi il est impossible qu'elles puissent former une étendue.[31]

The conclusion of Arnauld's line of argument is that extension cannot but be infinitely divisible, to which he appends a call for meditation on the limits of our mind. Bayle's treatment of the problem can be seen as an answer to this attitude, and his argument takes a different, but equally radical turn:

[31] *Art de Penser*, Part IV, chapter I, p. 297 ['nothing is clearer than this principle: Two entities of zero extension taken together still do not have any extension; that is to say, an extended whole has parts. Take any two of these parts which we assume to be indivisible. I ask whether the parts have extension. If they do not have extension, they have zero extension and the two taken together cannot have extension; if the indivisible parts have extension, they have parts and are hence divisible' (Dickoff and James, p. 299); in the 1685 translation: 'there is nothing more clear, then that two *nothings* of an Extent can [sic] form an Extent; and that every Extent has parts. Now taking two of these parts which are suppos'd to be indivisible, I demand whether they have extent or no? If they have, then they are divisible and have parts; if they have not, then they are Nothings of an Extent, and so it is impossible they can form an Extent' (p. 168)].

If Extension existed, it would be composed either of the Mathematical Points, or of Physical Points, or of parts divisible *in infinitum*: But it is not composed of Mathematical Points, or of Physical Points, or of parts divisible *in infinitum*. Therefore it doth not exist. There is no fault in this Syllogism; the Sophism *à non sufficienti enumeratione partium* is not in the Major; the Consequence is therefore necessary, provided the Minor be true.[32]

Hume's answer to Bayle is, one may add, that there is indeed the fallacy *a non sufficienti enumeratione partium* in the major premise of this syllogism. Extension may well be, indeed actually is, composed of unextended coloured and tangible points.

ALIQUOT AND PROPORTIONAL PARTS

So, according to Hume, the minima of sight are unextended coloured points. This is a central aspect of Hume's argument in section 1, as in the whole of Part 2. It is now useful to read again the first moves of Hume's treatment:

'tis universally allow'd, that the capacity of the mind is limited, and can never attain a full and adequate conception of infinity: And tho' it were not allow'd, 'twou'd be sufficiently evident from the plainest observation and experience. 'Tis also obvious, that whatever is capable of being divided *in infinitum*, must consist of an infinite number of parts, and that 'tis impossible to set any bounds to the number of parts, without setting bounds at the same time to the division. It requires scarce any induction to conclude from hence, that the *idea*, which we form of any finite quality, is not infinitely divisible, but that by proper distinctions and separations we may run up this idea to inferior ones, which will be perfectly simple and indivisible. In rejecting the infinite capacity of the mind, we suppose it may arrive at an end in the division of its ideas; nor are there any possible means of evading the evidence of this conclusion. (*T*/26–7)

In this passage there are still quite a few unsolved difficulties. But I am now in a position to resolve some of the problems: for instance, the function and meaning of the infinite number of parts of an infinitely divisible object.

I have indicated that, according to Flew, Hume's principle is false, because it implies a wrong notion of division and of parts. On the same problem, Laird presents a more specific argument. He maintains that the crucial objection to Hume's statement is that a division *in infinitum* only implies an infinity of proportional, not of aliquot, parts. This objection has a strong historiographic appeal, because it was familiar

[32] 'Zeno of Elea', rem. F, p. 3078.

to Hume himself – who, by the way, chooses to show his knowledge of it by rejecting it, in a footnote, as 'entirely frivolous'. Now Laird claims that this objection 'was the opinion of most good Newtonians', and that Hume rejects it 'very cavalierly', in a way which 'appears to be a mere *petitio*':[33]

It has been objected to me, that infinite divisibility supposes only an infinite number of *proportional* not of *aliquot* parts, and that an infinite number of proportional parts does not form an infinite extension. But this distinction is entirely frivolous. Whether these parts be call'd *aliquot* or *proportional*, they cannot be inferior to those minute parts we conceive; and therefore cannot form a lesser extension by their conjunction. (*T*/30 fn.1)

In my opinion, the footnote concerning the distinction between aliquot and proportional parts reveals some crucial features of Hume's argument concerning the doctrine of infinite divisibility. One may begin by considering what the terms 'aliquot' and 'proportional' mean in Hume's footnote. Laird, as I have just noted, attributes the distinction to 'most good Newtonians', and refers to Clarke's third letter to Leibniz, where he finds that the expression 'division into parts' implies 'a figurative abuse of the word, *parts*'. Reading the whole of Clarke's passage, though, shows that the situation is not quite as simple as that:

Infinite space is one, absolutely and essentially indivisible: and to suppose it parted, is a contradiction in terms; because there must be space in the partition itself; which is to suppose it parted, and yet not parted at the same time [and this, as Clarke reminds the reader, is Newton himself: see Schol. ad Defin. 8 in the *Principia*]. The immensity or omnipresence of God is no more a dividing of his substance into parts; than his duration, or continuance of existing, is a dividing of his existence into parts. There is no difficulty here, but what arises from the figurative abuse of the word, *parts*.

So, in context, it is clear that the 'figurative abuse of the word' has more to do with the theologically delicate problem of how to think and talk about God than with any definition of the distinction between aliquot and proportional parts. At any rate, after Kemp Smith it is commonplace, and in the absence of any other evidence indeed seems fairly reasonable, to believe that the source of Hume's footnote is to be found in Bayle's *Dictionary*. Here, in the Remark F to the article 'Zeno of Elea', we find that aliquot parts are 'parts of a certain magnitude, and of the same denomination' (p. 3079). Aliquot parts are, in other words, extended and homogeneous with the whole: lines for a line, surfaces for a surface, and so on; they may be said to be components of the whole. Proportional

[33] Laird, *Hume's Philosophy*, p. 67.

parts appear a page earlier, where we find that the distinction between aliquot and proportional parts is one among those with which 'the schoolmen have armed' the hypothesis of infinite divisibility, in order to have

Matter for talk upon a publick Disputation, that their Relations may not suffer the Disgrace of seeing them mute. A Father or a Brother go away better satisfy'd, when the Scholar distinguishes between a *Categorematical* infinite and a *Syncategorematical* one, betwixt the Parts *communicantes*, and *non communicantes*, *proportional* and *aliquot*, than if he had answered nothing. (p. 3078)

In Note E, Bayle had also extensively discussed Aristotle's distinction between virtual infinity, or infinity 'in power', and actual infinity, presenting it as a 'wretched answer' to one of Zeno's arguments for the impossibility of motion (p. 3076). In short, the distinctions between aliquot and proportional parts and between potential and actual infinity are not explained unambiguously, but simply dismissed by Bayle as typical instances of scholastic sophistry. Given this, one may wonder how the terms 'aliquot' and 'proportional' are generally used in texts of the period.

In mathematical writings, 'aliquot parts' are commonly defined not in contrast with proportional parts, but rather by reference to definition 1, Book V of Euclid's *Elements*: 'A magnitude is a part of a magnitude, the less of the greater, when it measures the greater.' In addition to being homogeneous with its whole, an aliquot part is defined as one which is in the whole an exact number of times. The term appears with this meaning in Barrow's *Mathematical Lectures*, in Lecture 19 (with a *lapsus* of the translator or printer):

an Aliquot Part according to Euclid's Sense and Definition. A Part is a Magnitude of a Magnitude, [sic] a less of a greater, when the less measures the greater.[34]

Similarly in Chambers' *Cyclopædia*, under the entry 'Part', we find:

Aliquot PART, is a quantity which, being repeated any number of times, becomes equal to an integer. Thus 6 is an *aliquot part* of 24; and 5 an *aliquot part* of 30, &c. See ALIQUOT and MULTIPLICATION.

Aliquant PART, is a quantity which, being repeated any number of times, becomes always either greater or less, than the whole . . .[35]

[34] Barrow, *Mathematical Lectures*, p. 361.

[35] *Cyclopædia: Or, an Universal Dictionary of Arts and Sciences;* . . . by E. Chambers, FRS . . . In two volumes. London 1738. Yet the same definition of aliquot as opposed to aliquant parts is to be found in J. Ralphson, *A Mathematical Dictionary*, London 1702, s.v. 'Aliquot parts', defined as 'parts that are contained a certain Number of Times in their whole; as 2 is contained thrice in 6', while an aliquant part (s.v.) is 'a part contained in its whole, a certain Number of Times, leaving a Remainder over and above, as 2 in 7'.

Something more directly related to my present case appears, concerning the term 'proportional', in a page of the *Art de Penser* that we have already cited above:

cette partie dont la petitesse nous est déja incomprehensible, contient encore un autre monde proportionel, et ainsi à l'infini, sans qu'on puisse trouver aucune qui n'ait autant de parties proportionelles que tout le monde, quelque étendue qu'on lui donne (pp. 296–7).[36]

The parts here mentioned are called *proportionelles*; with this term, the text gestures toward that infinity of smaller and smaller worlds, each imagined by analogy to the one we are acquainted with, and which we have found to represent Hume's polemical target. Also, the proportional parts it talks about are components of a whole, and each of them is a composite whole in turn, and so on. In other words, they seem to share a crucial feature of 'aliquot' parts. The distinction we are looking for, finally, does appear in Bayle's *Systeme Abregé de Philosophie*:

Les partes aliquotes sont celles qui répétées un certain nombre de fois, épuisent ou égalent le tout parfaitment, comme le pied par rapport à la perche qu'il égale, étant répeté six fois . . . Une partie qui ne fait pas cela, est appellée aliquante, comme cinq par rapport à quatorze . . . Les parties proportionelles sont celles qui décroissent dans une certaine proportion, comme lorsqu'on divise une quantité de deux pieds en deux pieds, le pied en deux demi-pieds, le demi pied en deux moitiez de demi pieds, &c. Ces parties ne son point égales entre elles, & n'ont point une grandeur certaine & determinée, comme les parties aliquotes, puis qu'elles peuvent s'appetisser & décrotîre à l'infini.[37]

Bayle concludes by reporting that, according to those who believe in the infinite divisibility of extension, it is the infinity of these proportional parts, contained in each body, that actually constitutes body: 'ceux qui veulent que [le continu] soit composé de parties divisibles à

[36] *Art de Penser*, Part IV, chapter I, pp. 296–7: see above, fn. 25.

[37] *Systeme Abregé de Philosophie*, in *Oeuvres diverses de Mr. Pierre Bayle*, vol. 4, A La Haye 1731 (reprint Hildesheim 1970), p. 293 ['Aliquot parts are those that, repeated a certain number of times, exhaust or equal the whole perfectly, as the foot is to the *perche*, which it equals when repeated six times . . . A part which does not do so is called an aliquant part, as for instance five with respect to fourteen . . . Proportional parts are those that decrease in a certain proportion, as when one divides a quantity of two foot into two feet, and the foot in two halves of a foot, and the half foot into two halves of a half of a foot, and so on. These parts are by no means equal to each other, and by no means have a certain and determined magnitude, as the aliquot parts do, since they can diminish and decrease to infinity.' My transl.].

l'infini' think that 'le nombre de ces parties est infini dans chaque corps'.[38]

The sample of texts briefly examined here has been chosen because it is reasonable to presume, for various reasons and with different degrees of confidence, that each of them was familiar to Hume. It is remarkable that none of them seems to lead me any further towards the solution of my problems. Even Bayle's *Systeme* defines proportional parts clearly and crisply only to conclude that they are supposed to make up a body. What emerges is, apparently, a general lack of terminological clarity, which is somehow reflected in Hume's text: for, while in the footnote in *T*/30 Hume seems simply to reproduce the attitude expressed in Bayle's 'Zeno', we have seen that he also talks of 'the distinct idea of these numbers and of their proportions'. At any rate, the distinction between aliquot and proportional parts, both as it is mentioned and employed (more than actually presented) by Bayle, and indeed in the confusing way it appears in other texts presumably familiar to Hume, suggests a use of 'parts' which has very little to do with Hume's discussion of unextended coloured points, and simply does not make sense as an objection to Hume's approach; and this is, after all, precisely what he says about it in his cavalier footnote. This point needs a little further explanation.

> Whatever is capable of being divided *in infinitum*, must consist of an infinite number of parts, and . . . 'tis impossible to set any bounds to the number of parts, without setting bounds at the same time to the division.

If we surrender for a moment to the temptation to use the language Hume is so explicit in rejecting, we may summarise what we have seen in this section by saying that the conception of 'proportional parts' is somehow parasitic on the indivisible image. As Newman puts it, Hume

> is not denying that we have an idea of infinity; rather, he is asserting that the idea of a potential infinity cannot suffice to support the implications of the thesis of infinite divisibility.[39]

The implication of Hume's treatment of indivisibles is that we cannot conceive parts which do not have at least one of the features of aliquot parts – being components of the whole that is divided – and at the same time at least one of the features that aliquot parts cannot have – being

[38] *Ibid.*, p. 292 ['those who want [the continuum] to be composed of infinitely divisible parts' think that 'the number of such parts is infinite in each body'].

[39] Newman, 'Space', p. 5.

unextended. Surely they are very paradoxical entities, and it is a particular consequence of their paradoxical character that makes it possible to use them as the model for indivisibles in general. This is what is going to be examined now.

LEVELS OF PERCEPTION AND THE ABSOLUTELY MINUTE

I have said that there is an image or some images corresponding to the distinct ideas of 'these numbers and of their different proportions'. The fact is that the images corresponding to these ideas are not in the same proportion to each other as the numbers themselves. They imply the existence of different levels of phenomena, whose imaginal representations are all shaped on the same model – the unextended minima afforded by sight. To the minimum of sight no unique phenomenal minimum corresponds in principle, except within the bounds of each phenomenal or perceptual level:

> A microscope or telescope, which renders [the minute parts of distant bodies] visible, produces not any new rays of light, but only spreads those, which always flow'd from them; and by that means both gives parts to impressions, which to the naked eye appear simple and uncompounded, and advances to a *minimum*, what was formerly imperceptible. (*T*/28)[40]

And again:

> The only defect of our senses is, that they give us a disproportion'd image of things, and represent as minute and uncompounded what is really great and compos'd of a vast number of parts . . . This however is certain, that we can form ideas, which shall be no greater than the smallest atom of the animal spirits of an insect a thousand times less than a mite: And we ought rather to conclude, that the difficulty lies in enlarging our conceptions so much as to form a just notion of a mite, or even of an insect a thousand times less than a mite. For in order to form a just notion of these animals, we must have a distinct idea representing every part of them. (T/28)

When the mind jumps from one level of perception to another, it has no means to comprehend the gap between the two levels without losing its imaginal consistency. However, on the basis of the indivisibles provided in sense-perception by sight, the mind moves easily amongst atoms as well as amongst minute insects and the objects of everyday experience. For it does not matter what an indivisible of sight represents; what

40 Cf. Berkeley, *New Theory of Vision*, section 85: 'A microscope brings us as it were into a new world', in which the usual connection between sight and touch is lost.

matters is that it is an unextended indivisible and, as such, it may be considered to have an absolute value. So visual perception is inadequate and misleading, yet perfectly suitable as a general model for the imaginal content of such conceptual constructions as the ten thousandth part of the grain of sand.

In fact, perception is not just perfectly suited to the purpose: it is the actual source of such an imaginal content. There is only one kind of mental 'images', and such images consist of units which are immediately present, but not specific to sight: being units, they reappear without any change in every level of perception.[41] In this way, indivisibles of sight are found to be 'adequate representations' of indivisibles in general. At the end of section 1, Hume talks about

> the error of the common opinion, that the capacity of the mind is limited on both sides, and that 'tis impossible for the imagination to form an adequate idea, of what goes beyond a certain degree of minuteness as well as of greatness. Nothing can be more minute, than some ideas, which we form in the fancy; and images, which appear to the senses; since they are ideas and images perfectly simple and indivisible. (*T*/28)

The minima are not alien entities at all. From experience we know both indivisible impressions and indivisible ideas. However paradoxical, minima really are very easily understood.

At this point, we may solve a further practical complication. In the course of his discussion of the standard of equality, Hume writes that the points are

> so minute and confounded with each other, that 'tis utterly impossible for the mind to compute their number. (*T*/45)

We have seen something similar in the discussion of the spot of ink:

> when it has encreas'd to such a degree as to be really extended, 'tis still difficult for the imagination to break it into its component parts, because of the uneasiness it finds in the conception of such a minute object as a single point. (*T*/42)

Kemp Smith comments that

> in thus speaking of the simples as 'confounded with each other' Hume is running counter to his avowed position, that immediate awareness is infallible, and that perceptions are in all respects precisely what they are experienced as being.

[41] Cf. Berkeley's version of this, in *Vision*, section 80: the *minimum visibile* – being a *minimum* – is the same for all creatures.

The reason for Hume's contradiction is, according to Kemp Smith, his commitment to the theory 'that it is in simples . . . that compounds exist'. More generally and radically, Maund presents the problem implicit in Hume's talk of the confusion of minima in the following terms: 'In Hume's philosophy, the notion of the fallibility of the senses should be regarded as meaningless.'[42]

Now, I suggest that one may formulate the problem as follows: when treating many perceptual units at once, the mind has to account for their number – that is, it has to deal at once with a whole and its parts, and is put in a situation where it has to consider two different perceptual levels together. It can do so only to a certain extent, and always with a feeling of uneasiness. This is the problem of extension in general, of the areas of geometrical figures (*T*/45) as well as of complex entities like the insect.

> We can form ideas, which shall be no greater than the smallest atom of the animal spirits of an insect a thousand times less than a mite;

but to have an adequate image of the whole insect is impossible 'by reason of the vast number and multiplicity of these parts' which, taken in isolation, are so easily conceived on the model afforded by visual indivisibles (*T*/28). Putting the matter in these terms makes it clear that neither the reference to the confusion of minima nor that to the 'disproportioned image of things' given by the senses implies that our senses are fallible; for the infallibility of the senses may still be found in the absolute minuteness of their unextended units. But, between the spot of ink that is indivisible just before disappearing, and the spot of ink that is extended and composed of parts, there is the gap that divides two different levels of perception. They are like two different worlds, and common experience happens to cut through different worlds in this way all the time. It is the way perception works. The object of natural philosophy is nothing that can be properly accounted for, or even made sense of, taken in isolation from the totality of human experience. The abstract consideration of sense-perception, though exemplary for, and an important stage of, the inquiry into human nature, does not in itself say much about human experience. Sense-perception is the safest guide; but, in order to receive from it suitable answers to our questions, it has to be considered within the totality of perception, together with acts of the judgement and emotions.

[42] Kemp Smith, *Philosophy*, pp. 278–9; Maund, *Knowledge*, p. 202; see also Laird, *Hume's Philosophy*, p. 68.

We are now at the major – and final – problem of the present discussion. As I have said, Hume seems to be in no position to conclude, from the limited capacity of the mind and the infinite number of parts in anything infinitely divisible, that

> the idea, which we form of any finite quality, is not infinitely divisible, but that by proper distinctions and separations we may run up this idea to inferior ones, which will be perfectly simple and indivisible. (*T*/27)

Now we know where Hume finds these indivisibles – they present themselves, as indivisible images, both to the imagination and to the senses. But one more general question still stands out concerning the refutation of the 'doctrine of infinite divisibility': why does Hume present the existence of indivisible images as incompatible with the 'doctrine of infinite divisibility'?

Let us recapitulate. I have shown that there are:

1. Different phenomenal or perceptual levels, like the flea on the flea in Swift's poem;
2. A relation, expressed in terms of numbers and their proportions, between those levels or fleas;
3. An imaginal apparatus which always employs exactly the same materials to represent to the mind each of the different levels or fleas.

So far, so good. However, it seems that, if the image representing the grain of sand is no different from those representing its thousandth and ten thousandth part, the 'distinct ideas of these numbers and of their different proportions' are all that is needed in order to enforce a 'doctrine of infinite divisibility' – even though one that differs from those of Arnauld or Malebranche.[43] Can we not move from level to level without end? Why does Hume consider the inability of the mind to form images in the same proportion to each other as the numbers as the crucial point of his discussion? In other words, it is not clear why the lack of a coherent imaginal content should be such a problem after all. What is it, then, that stops the process of divisions and subdivisions? It is possible, I suggest, to answer this question by saying that, from a certain point of view, the crucial role here is played by reality itself.

[43] Indeed, as Newman, 'Space', p. 5, correctly remarks, even the similarity between Hume's version of the limited capacity of the mind and Locke's description of infinity is suggestive of such an option.

TEMPTATIONS

The question of what it is that stops the division has been a recurrent one throughout my discussion. In the preceding pages, from time to time, I have been close to some possible answers, which have been implicitly rejected. Now, however, certain of these deserve further exploration.

A first possible answer to my crucial question is related to the notion of a point. I have shown that Hume distinguishes his points from both mathematical and physical points. It could hardly be more difficult to make sense of Hume's use of these terms. Proving the impossibility of mathematical points, one is told, is an absurdity (*T*/32–3); nevertheless, mathematical points are said to be non-entities, except if one considers them as coloured and tangible (*T*/40) – that is, it may be remarked, no longer just as mathematical points. Physical points, on the other hand, are absurd (*T*/40), but on occasion Hume calls the coloured points physical (*T*/112) – a notion which reappears in the only passage of the *Enquiry* devoted to the matter (*E*/124 fn.). It is tempting to maintain that behind this terrible confusion there is no real riddle: Hume's coloured and tangible points are indeed physical, and what Hume is presenting in the *Treatise* is, after all, a weirdly expressed but rather unsurprising form of physical atomism. This interpretation uses the *Enquiry*, the mature and recognised work, to cast light on the juvenile shortcomings of the *Treatise*. It has the crucial advantage of getting rid of the paradoxical character of Hume's coloured points, and affords an explanation of the rejection of the doctrine of infinite divisibility which is easily understood and consistent. Its appeal is strong and evident. Nevertheless, in my opinion, this way of reading the *Treatise* is a temptation which is better resisted – for it cuts the knots it is supposed to untie. For instance, it makes the case of the grain of sand utterly incomprehensible. It solves a problem raised by a text by making the text itself incredibly confused, indeed inconsistent; in addition, while eliminating the paradoxes involved in Hume's treatment of perceptual points, it also eliminates, most unfortunately, the really interesting and indeed important features of such a treatment – its stimulating ambiguities.

A more sophisticated version of this kind of interpretation may be produced by reference to the distinction between aliquot and proportional parts.[44] Why does Hume reject this distinction as frivolous? If

[44] I owe the following temptation – and a staunch defence of it – to Nick Jardine.

one calls an indivisible point an aliquot part, Hume's treatment of, say, the grain of sand can be transposed into the following terms: we can conceive as many divisions and subdivisions of the grain of sand, as we like; each of these divisions will uncover proportional parts, whose notion depends, on the one hand, on a relation, and, on the other, on the conception of some aliquot part. Aliquot parts would thus have the role of affording the imaginal content of proportional parts. Their conceptual inevitability would express the limitation of human mind in the most basic, and not unattractive, phenomenalistic terms: whatever process of division we perform, we shall as a matter of fact stop somewhere. From this point of view, one may safely suppose that Hume calls the distinction between proportional and aliquot parts frivolous because the two kinds of parts cannot be talked about separately; they always go together, as correlatives, in any conceivable process of division.

This reading is much more subtle than the first one. Instead of a physical atomism, it proposes a psychological one more congenial to Hume's approach. The mathematical notion of potential infinity is made dependent on the psychological notion of a point. One is left with a reasonable residue of inconsistencies in the text, and these inconsistencies, on the whole, look very much like marginal by-products of the (notorious) involution of the style of the *Treatise*. In short, this interpretation is a real temptation. However, I find it unconvincing – it does not feel right. For a start, the way the distinction between proportional and aliquot parts makes its appearance in the *Treatise* is, from the point of view of this reading, deeply unsatisfactory. Is it possible that, in order to solve one of the major problems of section 1, one has to resort to a distinction introduced only in passing, in a casual footnote in section 2, and then only to be rejected? But this is not all. This reading imposes on Hume's text a terminology that only apparently belongs to its frame of reference: for there is no clue in Hume's text as to what 'aliquot' and 'proportional' parts are supposed to be. So, if I decide to use these terms I must myself give them a meaning. And, at this point, the element of anachronism is not the worst problem with this reading. For my distinction between aliquot and proportional parts would necessarily be such as to appear, when placed in the context of the style of thought of the *Treatise*, much too specific – too directly borrowed from the language of mathematics, with a tinge of scholastic metaphysics if I am a bit more sophisticated – to provide the answer I am looking for. Is it really impossible to produce a more general, more convincing argument?

Of course, it is perfectly possible. Given the style of thought we are acquainted with in the *Treatise*, it is extremely tempting, and also quite consistent, to say that, since any process of division in practice stops somewhere, it is reasonableness that stops the possibility of an infinite divisibility. Once we have conceived a mite and then, through subsequent divisions, attained the notion of an insect a thousand times less than a mite, its animal spirits, and the smallest atom of its animal spirits; well, how much further would we push, and indeed how sensible would it be to try to provide the foundation for further divisions? The weakness of this interpretation is evident. The fact that it may appear pointless and unreasonable to keep dividing and subdividing the smallest atom of the animal spirits of an insect a thousand times less than a mite certainly does not mean that it is impossible, let alone that it is proved to be impossible in principle. And Hume is very explicit in stating that it is the positive demonstration against the infinite divisibility he is aiming at:

> I doubt not but it will readily be allow'd by the most obstinate defenders of the doctrine of infinite divisibility, that these arguments are difficulties, and that 'tis impossible to give any answer to them which will be perfectly clear and satisfactory. But here we may observe, that nothing can be more absurd, than this custom of calling a difficulty what pretends to be a demonstration, and endeavouring by that means to elude its force and evidence. (*T*/31)

Nevertheless, this interpretation looks promising. Is it possible to make it stronger?

For a start, it can be made more convincing by following up a suggestion from another part of Hume's text.

> A musician finding his ear become every day more delicate, and correcting himself by reflection and attention, proceeds with the same act of the mind, even when the subject fails him, and entertains a notion of a compleat *tierce* or *octave*, without being able to tell whence he derives his standard. A painter forms the same fiction with regard to colours. A mechanic with regard to motion. To the one *light* and *shade*; to the other *swift* and *slow* are imagin'd to be capable of an exact comparison and equality beyond the judgments of the senses. (*T*/48–9)

In the same way, once we are used to comparing the sizes of figures with increasing exactness, first by their appearance, then by a more careful review and reflection, then by juxtaposition or by a common measure, and so on, we

> suppose some imaginary standard of equality, by which the appearances and measuring are exactly corrected, and the figures reduc'd entirely to that proportion. (*T*/48)

It is in this way also that we are led from the coherence of the appearances of objects to the senses to the supposition of the continued existence of objects:

as the mind is once in the train of observing an uniformity among objects, it naturally continues, till it renders the uniformity as compleat as possible. The simple supposition of their continu'd existence suffices for this purpose, and gives us a notion of a much greater regularity among objects, than what they have when we look no farther than our senses. (*T*/198)

This feigning of a perfect standard of tone, light, velocity, equality, and uniformity is actually an expression of a very general tendency of the mind; nothing, says Hume, is

more usual, than for the mind to proceed after this manner with any action, even after the reason has ceas'd, which first determin'd it to begin. (T/48)

It is a common feature of the imagination in its activity – which Hume describes using once more an exemplary natural-philosophical metaphor:

The imagination, when set into any train of thinking, is apt to continue, even when its object fails it, and like a galley put in motion by the oars, carries on its course without any new impulse. (T/198)

This sort of mental inertia accounts for the imaginary standards of tone, light, velocity, equality, and uniformity; and, given the generality Hume appears to attribute to it, it is also very easily applied to explain why it is so difficult for us to give up the idea of an infinite divisibility. For, once we know that we can conceive smaller and smaller parts of a grain of sand, by reference on the one hand, to the relation of proportion of quantity and number, and, on the other, to our invariable imaginal apparatus, we are naturally inclined to continue 'like a galley put in motion by the oars', that is, to feign the possibility of an infinity of levels each infinitely smaller than its predecessor. But this fiction, however natural, is 'useless as well as incomprehensible' (*T*/48); we do not need it for our operations, since each of them does stop somewhere anyway, and we cannot comprehend it, since our mind is finite. The crucial fact is that Hume seems to define and measure 'possible' and 'impossible' in terms of the concrete capacity of human mind,[45] and the results are somewhat unexpected. The key-point in the discussion of infinite

[45] The problem of the use and meaning of these and related notions (for instance, 'absurd': see below, fn. 50) in Hume's texts has been studied extensively in connection with the analytic/synthetic distinction: for a detailed treatment of this question see below, chapter 3, 'Demonstrative knowledge', and the bibliography quoted there.

divisibility may now be put as follows: it is at once perfectly natural and utterly unjustified that we feign an operation that is only defined by our way of performing it, and then pushed beyond our capacity ever to perform it. Hume's arguments appear suddenly very tight. However, in order to give them a fuller meaning, and to explain why Hume claims demonstrative force for them, something more has to be added. It is now time to discuss the role of the presupposition about external reality underlying Hume's discussion of the doctrine of infinite divisibility.

REALITY ITSELF

Let us go back to the grain of sand: I said at the outset that the main problem in section 1 is the adequacy of our perception. Adequacy to what? It is the mention of 'the thing', the grain of sand meant, apparently, as an external object, that first shows how Hume's discussion involves some sort of presupposition about the existence of external reality. Such a presupposition is not developed, in fact it is not even enunciated, but simply underlies Hume's discussion line after line. It will only appear clearly, but again without being made a problem to be discussed, in section 2, where Hume states that

> wherever ideas are adequate representations of objects, the relations, contradictions and agreements of the ideas are all applicable to the objects,

so that, since

> our ideas are adequate representations of the most minute parts of extension . . . whatever *appears* impossible and contradictory upon the comparison of these ideas, must be *really* impossible and contradictory, without any further excuse or evasion. (*T*/29)

It is after this direct reference to reality that Hume goes on to claim the status of demonstration for his arguments against the doctrine of infinite divisibility. So one may well suspect that this reference has a crucial role within his line of argument. To start with, one may note that it is still possible, after Hume's perceptual analysis, to consider infinite divisibility as exposing the limitation of human mind with respect to something – the real world, out there – that by definition surpasses human perceptual capacity. In other words, if Hume's rejection of infinite divisibility is to be considered a full demonstration against infinite divisibility, instead of simply a more sophisticated restatement of the limited capacity of the mind, it is necessary to consider what role a reference to external reality can play in the matter. I shall show that Hume's realistic presupposition is at once the key-stone of his argument and an answer to the problem of establishing how far a realistic presupposition can go in an argument for or against infinite divisibility – and, more generally, of how the relationship between perception and the real world out there is established.

It may be useful, for a start, to consider the realistic implications latent in section 1. It is evident that talking about spots of ink, grains of sand, mites, and so on, in the way Hume does, implies some non-phenomenalistic assumptions about spots of ink, grains of sand, insects, and other external objects. As a phenomenalist interpreter wonders, 'What is *the* "spot" if the "impressions" form a series?'[46] The same may be said about the grain of sand and the mite: without a presupposition of some kind about external objects, the perceived objects discussed by Hume disintegrate into clusters of unrelated sense-impressions, ideas, and emotions. However, to make sense of Hume's allusion to the 'thing' in this context, as well as in other passages, it is not at all necessary to consider his realistic presupposition to be very strong. Perception presents 'objects', not scattered data, and it is objects that constitute the original materials of our discussion in Hume's phenomenology of human experience:

> 'tis certain, that almost all mankind, and even philosophers themselves, for the greatest part of their lives, take their perceptions to be their only objects, and suppose, that the very being, which is intimately present to the mind, is the real body or material existence. (*T*/206; see also *T*/205 and cf. *T*/219)

Only on reflection do we dissolve perceived objects into perceptual data which seem not consistent with each other: the figure of objects varies all the time, as well as their size, their colour, and their other sensible qualities (*T*/211). In the same way, we discover that perception presents us as indivisibles objects that in a different situation are composed of parts. The organisation of these scattered data, the establishment of their relations to each other, the invention of their consistency through the construction of the idea of an object, are operations that happen before any reflection. The spot of ink, the grain of sand, the mite Hume talks about in section 1, are familiar objects, easily recognised and readily classified. It is our reflection that makes them look unfamiliar and resolves them into clusters of different, indeed inconsis-

[46] Laird, *Hume's Philosophy*, p. 68

tent, perceptual data. Realism is, in this sense, the expression of a natural, and as such an extremely powerful, tendency of the mind:

nature is obstinate, and will not quit the field, however strongly attack'd by reason. (*T*/215)

But, in this way, it also becomes clear that this kind of realism is, more than a problem for, or a notion in, Hume's philosophy, one of the characters that appear on the stage of the *Treatise*, together with scepticism (in particular in its phenomenalistic variety).[47] This is a crucial point, and it needs more detailed discussion.

The first explicit, if brief, treatment of the idea of the existence of external objects – that is, the first recognition that it is a matter which needs considering – appears at the end of Part 2, in the last paragraphs of section 6, 'Of the idea of existence, and of external existence'. The conclusion is that:

The farthest we can go towards a conception of external objects, when suppos'd *specifically* different from our perceptions, is to form a relative idea of them, without pretending to comprehend the related objects. Generally speaking we do not suppose them specifically different; but only attribute to them different relations, connexions and durations. (*T*/68; cf. 67)

So, in principle, it is possible to suppose absolute space, say, as specifically different, that is different in character, from our spatial perceptions, which only constitute relative space. On the other hand, referring later to the passage just quoted, Hume writes:

as to the notion of external existence, when taken for something specifically different from our perceptions, we have already shewn its absurdity. (*T*/188)

At first sight, this may look like a pretty clear condemnation of any realistic interpretation. It is, however, easy to maintain that this is not the case, emphasising the fact that what is called 'absurd' is the supposition of a difference in character between external reality and perception, not of the existence of external reality as such: external objects are always supposed to exist, and this cannot even be made a subject for discussion.[48] What is now more important, though, is that the 'absurdity' in the quotation cannot in any case be used to support a purely

[47] Cf. Livingston, 'Sellarsian Hume', who considers Part 4, section 2, 'Of scepticism with regard to the senses' to have 'the form of a dialectical drama' (p. 284).

[48] See Yolton, *Perceptual Acquaintance*, p. 149. To put it in Yolton's terms – in fact Hume's: see *T*/257 – what Hume calls absurd is the supposition of a specific, not of a numerical difference.

phenomenalistic reading of Hume's position. For the term 'absurd' is commonly employed in the *Treatise* not in a logical, but in a psychological sense, that is as a synonym of 'inconceivable', or even of 'unreasonable'.[49] Thus, what Hume is saying when he describes 'the notion of external existence, etc.' as 'absurd' is that we cannot suppose that we can form an adequate notion of external existence – external existence is, strictly speaking, inconceivable. That is, he is simply repeating once more what he has been saying all along: that the supposition of an external existence is no metaphysical principle and cannot lead our discussion any farther. Moreover, Hume is also quite clear in his rejection of any purely phenomenalistic position, which he calls 'the most extravagant scepticism' (*T*/228). Phenomenalism – or, to use Hume's words, 'rejecting the opinion of a continu'd existence' –

> has been peculiar to a few extravagant sceptics; who after all maintain'd that opinion in words only, and were never able to bring themselves sincerely to believe it. (*T*/214)

On the other hand, phenomenalism, being a variety of scepticism, is, in turn, just like realism, the expression of a natural tendency – it 'arises naturally from a profound and intense reflection on those subjects' (*T*/218) and is the natural reaction of the mind when faced with the depth of philosophical discussion. Thus it is

> a malady, which can never be radically cur'd, but must return upon us every moment, however we may chace it away, and sometimes may seem entirely free from it. (*T*/218)

So, to summarise, both realism and phenomenalism are natural, and yet both, when transformed into consistent metaphysical positions, are found wanting in one specific and, for Hume, crucial respect: both are unreasonable. Whether the world out there exists is no matter for metaphysical discussion; out of a natural and compelling attitude of the mind, we assume it anyway. Given this, not surprisingly it is 'carelessness and inattention' that give the solution we cannot obtain through any metaphysics:

> the sceptic . . . must assent to the principle concerning the existence of body,

[49] See, for instance, *T*/40 and *T*/50. This use of 'absurd' is noted in passing in this connection by Strawson, *Secret Connexion*, p. 52: Hume 'says that the notion of external objects specifically different from perceptions is "absurd". But he does not really mean what we mean by "absurd".' This remark has given rise to the inappropriate sarcasm of Winkler, 'The new Hume', p. 554, fn. 7. See fn. 45 above.

tho' he cannot pretend by any arguments of philosophy to maintain its veracity: (*T*/187)

and

Carelessness and in-attention alone can afford us any remedy. For this reason I rely entirely upon them; and take it for granted, whatever may be the reader's opinion at this present moment, that an hour hence he will be persuaded there is both an external and internal world. (*T*/218)[50]

The problem of the existence of external objects is nothing but the problem of our propensities, tendencies, natural beliefs. Tendencies and propensities guide the whole discussion of external objects in Part 4: the imagination – as we have seen already – tends to let all its operations go too far through a sort of mental inertia (*T*/198–9); the mind tends to mistake two ideas associated by a relation (*T*/202); perplexities and feelings of unease in the face of contradictions produce the propensity to feign the continued existence of objects (*T*/205), and it is the propensity of imagination that determines our belief in it (*T*/208). In the end, we find that the entire metaphysical discussion has been referred to concrete human experience, and that metaphysics itself, with all its excesses, appears as nothing but a response to a natural tendency of the mind. This is also true if we consider the problem from another point of view. Hume objects to 'modern philosophy' because of the distinction between primary and secondary qualities:

if colours, sounds, tastes, and smells be merely perceptions, nothing we can conceive is possest of a real, continu'd, and independent existence; not even motion, extension and solidity, which are the primary qualities chiefly insisted on. (*T*/228)

It is secondary qualities which, however variable and apparently unreliable, determine the texture of human experience. The discovery that they, like vice and virtue, 'are not qualities in objects, but perceptions in the mind', is certainly 'a considerable advancement of the speculative sciences', but 'it has little or no influence on practice' (*T*/469). They inform all human life. Moreover, the only way we can find a foundation for our conception of the external objects we suppose to exist is through the most perception-dependent among their qualities: colour, to which the primary qualities of motion, extension, solidity are inevitably reduced.

50 See T/211, 216–17. It may be interesting to remember that, as R. Fogelin points out in his *Philosophical Interpretations*, p. 11, fn. 5, in the eighteenth century 'carelessness' may refer to 'freedom from care' rather than, or at least as well as, to 'looseness of argumentation' (see S. Johnson, *A Dictionary of the English Language*, London 1755, entry 'careless', n.2).

So, metaphysics responds to a natural tendency of the mind; realism and phenomenalism coexist and are staged as the symmetrical expressions of its (natural) excesses. What space is still left, after this, for a reasonable, that is a justified and balanced, consideration of the relationship between perception and external reality? At the beginning of this section we have seen that, according to Hume, since

> our ideas are adequate representations of the most minute parts of extension . . . whatever *appears* impossible and contradictory upon the comparison of these ideas, must be *really* impossible and contradictory. (*T*/29)

The ideas of extensionless, that is, absolutely minute, indivisibles are adequate; thus what is contradictory in the consideration of them is smoothly transferred to reality. The passage from perception to reality is made possible by the adequacy of perception. In Part 4, section 2, we find a more complete – and more difficult, more subtle – statement of the relation Hume establishes between perception and the suppositions about external objects:

> As every idea is deriv'd from a preceding perception, 'tis impossible our idea of a perception, and that of an object or external existence can ever represent what are specifically different from each other. Whatever difference we may suppose betwixt them, 'tis still incomprehensible to us; and we are oblig'd either to conceive an external object merely as a relation without a relative, or to make it the very same with a perception or impression. (T/241)

So, we can perfectly well suppose that there is some specific difference between external objects and perceptions – however, to this difference no particular content can be given:

> any conclusion we form concerning the connexion and repugnance of impressions, will not be known certainly to be applicable to objects; but . . . on the other hand, whatever conclusions of this kind we form concerning objects, will most certainly be applicable to impressions . . . when we first form our reasoning concerning the object, 'tis beyond doubt, that the same reasoning must extend to the impression: And that because the quality of the object, upon which the argument is founded, must at least be conceiv'd by the mind; and cou'd not be conceiv'd, unless it were common to an impression. (*T*/241–2)

With this, Hume has sketched very neatly and clearly the key-feature of his attitude concerning the problem of external objects. Since it is possible to suppose the existence of an indeterminable difference between our perceptions and external objects, it is not safe, in principle, to transfer to external objects what we discover about our perceptions. On the other hand, we may suppose that reality is different from perception; but whatever we imagine about it must be applicable, in principle, to perception. In this way, a constraint is imposed on our suppositions: it excludes the possibility of external objects of which we cannot make sense, even in the most general way, in our own terms – that is, in terms of perception. So, perception is the general criterion and the crucial test for our suppositions about external objects: in principle any incompatibility between such suppositions and perception is to be considered as a sign that something has gone wrong in our reflections. Reality does indeed exist out there – we cannot help supposing it – as a 'something' whose features are unknown, except in one respect: we cannot even imagine them as incompatible with perception in general. On this single point, however, we can rely with full, absolute confidence. And this means that, with such a constraint, our suppositions about external objects can, without losing their indeterminacy, play some role in a discourse about human nature and even contribute to the investigation of perceptions.

This brings us back to the discussion of infinite divisibility – to the grain of sand, again. I have said that a natural realistic presupposition is not only actually present in Hume's discussion; it is also a condition for Hume's rejection of infinite divisibility to have the required – that is, demonstrative – strength. This is evident from the way Hume presents the matter in section 2. What specific part does this presupposition play? How does it affect, for instance, the notion of perceptual points?

> 'tis not for want of rays of light striking on our eyes, that the minute parts of distant bodies convey not any sensible impression; but because they are remov'd beyond that distance, at which their impressions were reduc'd to a *minimum*, and were incapable of any farther diminution. A microscope or telescope, which renders them visible, produces not any new rays of light, but only spreads those, which always flow'd from them; and by that means both gives parts to impressions, which to the naked eye appear simple and uncompounded, and advances to a *minimum*, what was formerly imperceptible. (*T*/27–8)

Microscopes and telescopes discover new levels of perception, new worlds: what is it that they actually do? They spread rays of light. How far can they go? Hume's argument seems to imply a finite and fixed number of rays of light flowing from each visible body. Actually, in this passage Hume is speaking the language of a kind of natural philosophy textbook, quite popular in the 1720s and 30s, where light was conceived as a finite number of minute material particles emitted at high speed by visible bodies.[51] For instance, it is the notion of light to be found in George Cheyne's *Philosophical Principles of Natural Religion*.[52] The emphasis is on the materiality, extreme minuteness, and exceeding speed of the particles of light. In Benjamin Martin's *Philosophical Grammar*, a manual published in 1735 which is a classic of eighteenth-century popularisation, the minuteness of the particles of light is reinforced with an argument borrowed from Bernard Nieuwentijtd:

It is computed, that in a Second of a Minute, there flies out of a burning Candle, the following Number of Particles of Light, 418660000000000000000000000000000000000000000 [in Nieuwentijtd: 418660×10^{39}], which is 10000000000 or Ten Millions of Millions Times a bigger Number than 10000000000000000000000000000000000, the Number of the Grains of Sand, computed to be contained in the whole Earth.[53]

It is evident that there is here yet another interpretative temptation. It is true that in books like Martin's and Cheyne's this conception of light coexists with a doctrine of infinite divisibility.[54] But, given the context created by Hume's discussion, it is very tempting to read its appearance in the argument concerning microscopes and telescopes in the following way: using microscopes and telescopes one would, beyond a certain level of magnification, simply and literally run out of light – that is, one would reach a degree of minuteness coinciding with the very limit of the physical world. With this, Hume's rejection of infinite divisibility appears complete and quite powerful. The problem with this reading is that it pushes Hume's discussion far into natural philosophy – a bit too far for Hume's standards in this respect. Worse, it commits Hume to a very precise description of the reality of rays of light, in a way that his explicit stand on the problem of external objects does not seem to allow.

51 See G. Cantor, *Optics after Newton. Theories of Light in Britain and Ireland 1704–1840*, Manchester 1983, esp. pp. 32ff.

52 G. Cheyne, *Philosophical Principles of Natural Religion: Containing the Elements of Natural Philosophy and the Proofs for Natural Religion, Arising from Them*, London 1705. See in particular chapter 1, sections 30–6, pp. 60ff.

53 B. Martin, *The Philosophical Grammar; Being a View of the Present State of Experimental Physiology, or Natural Philosophy*, London 1735, p. 60. The amazing computation is in B. Nieuwentijtd, *The Religious Philosopher: or, the Right Use of Contemplating the Works of the Creator*, London 1718. This work, originally published in Dutch, Amsterdam 1714, was translated into English by J. Chamberlayne and reprinted several times. I have consulted the 3rd edition, London 1724, where the computation is in vol. 2, sections 13–16, pp. 454–60; section 17, pp. 460–1, compares the number of the particles of light emitted by the candle with the number of grains of sand on Earth.

54 See Martin, Part 1, chapter 3, pp. 40ff., 'Of the Divisibility of Matter, of the Infinity thereof. . .'; the whole Part 2, first published in 1733, of Cheyne's *Principles* is on the 'Nature and Kinds of Infinites, their Arithmetick and Uses, and the Philosophical Principles of Revealed Religion'.

However, once one has overcome this final temptation, one thing at least is made completely clear by the argument on microscopes and telescopes: the different worlds discovered by microscopes and telescopes are, in fact, not different at all – they are again, all of them, other perspectives on exactly the same 'world' that our own eyes keep discovering all the time. The notion of its real existence, out there, is a general and undetermined propensity of the mind. And this is enough to round off Hume's argument against infinite divisibility.

It is now time to give my final exposition of the case of the grain of sand. We have an image of the grain of sand – distinct from our idea of it – which is indivisible. Nevertheless, we can easily conceive of dividing the grain of sand into a thousand or ten thousand parts, since our notion of these numbers and of their proportions is independent of our imaginal apparatus. This means that each of these parts will be represented by the imagination through the same indivisible image representing the grain of sand itself; with the division, we consider the grain of sand so to speak in a different light and move to a different perceptual level, which is as conceivable as the first one, where what is indivisible is not the grain of sand, but its thousandth (or ten thousandth) part. The process of division and subdivision may be described as jumping from one level to another much smaller, and so on. This process may be carried on as long as one likes but, obviously enough, is bound to stop somewhere, sooner or later. The imagination, however, tends to overlook this, and to follow its propensity to feign an endless process – an infinity of levels. This is a natural fiction without any grounds whatever, either in our practices or, for that matter, in the (finite) ability of comprehension of the mind. This is not all. The fact that no process of division is infinite also says something about external reality. Since whatever we suppose about reality must be applicable to perception, this feature of perception – its always presenting indivisibles, sooner or later – places a limit on reality itself. Reality cannot here go against perception. On this issue, perception and reality as it were touch each other and are fused.

The strength of Hume's argument – and the reason why it cannot be described simply as a 'physical atomism' – is that, while he says that any process of division must stop somewhere, he does not commit himself to say where. The minimal presupposition about the existence of external reality that I have described is the real pillar of his argument – and it is a very good pillar indeed. But, consistent with his statement that the existence of external reality cannot be treated metaphysically – "tis in vain to ask, *Whether there be body or not?*', because it 'is a point, which we must take for granted in all our reasoning' (*T*/187) – he sticks to his 'carelessness and in-attention', this time in a very specific and fortunate way. He maintains that we can perceive – that is, feel, imagine, think of – indivisible minima; however, he is very clear in stating that what is specifically represented by these minima of feeling and thought may not be, and indeed often is known not to be, indivisible. Perception is organised in such a way, that it cannot give any information about real indivisibles and where to find them. It is, rather, the very existence of perceptual minima that says something about reality: no matter where perception finds them, the minima it finds are absolute and, as such, represent not only their specific object, but also, even though in a different, more abstract, way, the unidentified and unidentifiable minima of reality. This may be not much, but it is the best we can rely on. With his typical, in this case so appropriate, 'stubborn vagueness',[55] Hume does not go beyond this.

55 The expression is in Laird, *Hume's Philosophy*, p. 37, about 'Hume's airy observation that somehow . . . simple ideas were derived from simple impressions'.

CHAPTER 2

A bundle of (organised) perceptions: Section 3 on the origin of the idea of space

> The essence of the mind being equally unknown to us with that of external bodies, it must be equally impossible to form any notion of its powers and qualities otherwise than from careful and exact experiments, and the observation of those particular effects, which result from its different circumstances and situations.
>
> (*T*/xvii)

INTRODUCTION

Chapter 1 has been devoted to Hume's arguments against the doctrine of the infinite divisibility of space and the idea of space in Book 1, Part 2, sections 1 and 2 of the *Treatise*; its main concern was the mind's capacity to establish and handle notions of external objects on the basis of its own perceptions. Now I intend to examine the origin of the idea of space, which is the subject of Book 1, Part 2, section 3 of the *Treatise*. Here, in about six pages, Hume shows how the idea of space derives from sense-experience and is the reflex in thought of the 'manner of appearance' of visual and tactile impressions, that is, of the way such impressions are arranged in sense-perception. The main focus of the following analysis is on the trace of the mind's activity in this 'manner of appearance'.

> Hume's entire philosophy was a philosophy of 'manners' as much as of 'impressions' and of 'ideas',

is the comment of John Laird in his old, excellent book on Hume. He goes on to remark, on the other hand, that concerning the idea of space Hume's expression 'disposition in a certain order' is possibly 'an echo of Cartesianism', and that

> Hume's explanation of this 'order' or 'disposition' was only to mention it . . . He was bound to hold that space was not a separate impression. For the same

reason he was bound to hold that no idea could copy it. And therefore he had no way of distinguishing this 'manner' or 'order' in which the individual minima were 'disposed' from any other manner or order, or, for that matter, from disorder. Yet the manner or order was the essence of space.[1]

Laird's remarks on this point are sensitive. In the literature, however, it is more common to find open reservations about the consistency of Hume's position on the origin of the idea of space. Typically, it is observed that, while Hume introduces his discussion of the origin of the idea of space with a reference to his 'first principle' that all our ideas make their first appearance to the mind in a corresponding impression, the actual treatment of the matter seems to go in an entirely different direction – the idea of space, in so far as it derives from the manner of appearance of impressions, does not make its first appearance to the mind in any separate corresponding impression. Thus, Hume seems to many of his scholars to be confused or inconsistent. The treatment of the idea of space is, many suggest, a sign either of a fundamental inadequacy in Hume's metaphysical principles and approach or, at best, of an interesting, but confused and entirely unresolved, tension within his thought. For instance, Kemp Smith says that the study of the origin of the idea of space

opens with a recapitulation of the principles insisted upon in the introductory sections of the *Treatise*, and it closes with teaching out of keeping with these principles.

According to Kemp Smith, it would have been much easier for Hume to maintain that 'extensity is a feature of certain of our sensations'; in his opinion, the influence of Hutcheson explains why Hume does not opt for such a solution, but instead treats space and time as 'non-sensational'.[2] Hendel remarks that

the tenor of this study of space and time seems to be that our ideas of these realities are in the same situation with the idea of causal necessity, which we will describe later as the feeling of our mental determination or disposition. Yet Hume does not propose this conclusion at present . . . the spatial and temporal disposition of the objects is the *mind's* disposition of them as it perceives. As a relation, then, the orders would *subsist* in the mind but not be represented to the mind. This is what the argument suggests,

[1] J. Laird, *Hume's Philosophy of Human Nature*, London 1932, pp. 129 and 71.

[2] N. Kemp Smith, *The Philosophy of David Hume*, London 1941, pp. 273 and 280. The chapter devoted to Hume's treatment of space and time is actually entitled 'Hume's version of Hutcheson's teaching that space and time are non-sensational'.

even if it is not what Hume actually says; at any rate,

> this second Part of the *Treatise* is marred by the haste with which Hume rushed into publication . . . the discussion bears all the marks of belonging to that very early period of thought which followed upon his discovery at the age of eighteen.[3]

Now, the oddity – to say the least – of Hume's line of argument concerning the origin of the idea of space is evident. It becomes a very interesting oddity, however, if one notices that there are several other cases in which Hume's use of his first principle has the same puzzling quality. The same way of arguing also appears, for instance, in connection with abstract ideas, with belief, and with the self. In my opinion, given the number and the importance of the cases in which Hume refers to the first principle while introducing arguments which seem not to be in line with it, or even to violate it, it is unwise to dismiss the problem by pointing out Hume's inconsistency – for it seems likely that there is here an opportunity to gain a deeper understanding of the *Treatise*.

The aim of this chapter is to suggest an interpretation of Hume's treatment of the origin of the idea of space which is guided by, and focussed on, these oddities and puzzles. Its intended result is that, instead of too patent inconsistencies, there will appear more reasonable and more human shortcomings in Hume's text – some gaps, something left between the lines, some difficulties solved only in part, some answers just hinted at, a generally tortuous way of arguing. Some of the most striking of the apparent contradictions and inconsistencies commonly pointed out in the literature will no longer seem to be such, and

[3] C. W. Hendel, *Studies in the Philosophy of David Hume*, Princeton 1925, pp. 152 and 164. (For a line similar to Hendel's, see B. Mijuskovic, 'Hume on space (and time)', *Journal of the History of Philosophy* 6 (4), 1977, pp. 387–94). See also C. Maund, *Hume's Theory of Knowledge*, London 1937, p. 207, where one finds that the expression 'manner of appearance' 'only disguises [Hume's] confusion'; A. Flew, 'Infinite divisibility in Hume's *Treatise*', in D. W. Livingston and J. T. King (eds.), *Hume: A Re-evaluation*, New York 1976, p. 262, writes that in section 3 'Hume attempts . . . the uphill task of applying in this pair of recalcitrant cases [space and time] his own doctrine "that every idea, with which the mind is furnish'd, first makes its appearance in a correspondent impression" (*T*/33)'; and, according to J. Noxon, *Hume's Philosophical Development*, Oxford 1973, p. 115, 'Hume's unavailing struggles with space and time, his attempt to derive their idea from experience conformably with his first principle . . . yielded the least-admired part of the *Treatise*'. R. Kuhns, 'Hume's republic and the universe of Newton', in P. Gay (ed.), *Eighteenth Century Studies Presented to Arthur M. Wilson*, New York 1972, p. 80, says that space and time 'constitute (though Hume never explicitly recognizes this) important exceptions to the principle of verification by reference to impressions and. . . are "known" without the mediation of impressions of themselves . . . In offering this "solution", Hume has denied both the mathematicians and his own epistemology.'

some others will be read as adequate expressions of fundamental metaphysical problems. I shall maintain that it is possible to find good reasons for Hume to treat the origin of the idea of space and the notion of 'manner of appearance' in the way he does. More particularly, I intend to show that, in treating the origin of the idea of space, the first principle is applied consistently, if not exactly in the way one might expect, and that the peculiarities thus discovered in the idea of space are to be referred to the special status of that idea. My suggestion is that section 3 makes good sense once one appreciates the original contribution of the self to the origin of the idea of space: the idea of space is the expression of a quality of the mind which, as such, can be defined only negatively through the first principle.

On the basis of my reading of the pages on the origin of the idea of space, something will become more clear about the possible uses of the principle of the correspondence between impressions and ideas as a tool for the metaphysical examination of mental contents. Also, I shall show something about the self – at once the most eminent victim and the absolute protagonist of Hume's 'science of human nature' – as it is dissolved and reconstructed by Hume's critique of experience.

IMPRESSIONS, IDEAS, AND THE ORIGIN OF THE IDEA OF SPACE

The most striking of all the apparent inconsistencies in the six pages on the origin of the idea of space is, of course, the one I have mentioned already – the one which opposes the idea of space to the very first principle of Hume's metaphysics, the principle of the correspondence of impressions and ideas. This is how this contradiction is first introduced:

> No discovery cou'd have been made more happily for deciding all controversies concerning ideas, than that above-mention'd, that impressions always take the precedency of them, and that every idea, with which the imagination is furnish'd, first makes its appearance in a correspondent impression. These latter perceptions are all so clear and evident, that they admit of no controversy; tho' many of our ideas are so obscure, that 'tis almost impossible even for the mind, which forms them, to tell exactly their nature and composition. Let us apply this principle, in order to discover farther the nature of our ideas of space and time . . . The table before me is alone sufficient by its view to give me the idea of extension. This idea, then, is borrow'd from, and represents some impression, which this moment appears to the senses. But my senses convey to me only the impressions of colour'd points, *dispos'd in a certain manner* . . . we may conclude with certainty, that the idea of extension is nothing but a

copy of these colour'd points, *and of the manner of their appearance.* (*T*/33–4, italics mine)[4]

There is nothing problematic with the coloured points: they are impressions, and they are duly copied by ideas. But what about their being 'dispos'd in a certain manner'? This 'manner of their appearance' or 'disposition' of the coloured points is a very ambiguous notion. It is certainly not an independent impression (Hume is explicit about this) – but then, what kind of impression is it supposed to be, and in what sense, on the basis of the copy principle, can it give rise to an independent idea, such as the idea of space? If the idea of space is not a copy of an independent impression, is it still compatible with, or indeed an application of, the copy principle? The appearance of inconsistency on this point is so open, so direct, that I suggest we start with the most basic interpretative doubts: perhaps there is something in the whole argument of section 3 that we are missing, and perhaps the role of the copy principle here is different from what we expected. Let us now review the whole argument step by step, starting with a careful look at the way Hume introduces the first principle in the first section of the *Treatise*.

All our simple ideas proceed either mediately or immediately, from their correspondent impressions. This then is the first principle I establish in the science of human nature . . . in order to prove the ideas of extension and colour not to be innate, philosophers do nothing but shew, that they are conveyed by our senses. To prove the ideas of passion and desire not to be innate, they observe that we have a preceeding experience of these emotions in ourselves. Now, if we carefully examine these arguments, we shall find that they prove nothing but that ideas are preceded by other more lively perceptions, from which they are derived, and which they represent. (*T*/7)

The principle of the correspondence between impressions and ideas is here illustrated with a reference to, among other things, the origin of the idea of extension; and this is also the first appearance of the idea of extension in the *Treatise* – figuring, together with the secondary quality of colour and just before passions and desire, as a cursory if significant illustration of the first principle. It is, in short, Hume's first-approximation formulation of both the first principle and his opinion about the origin of the idea of space. And the idea of space is so much in line with the first principle, that it is taken as an illustration, rather than simply an application.

[4] The significance of the fact that 'space' and 'extension' are used as synonyms is considered extensively in chapter 4.

The reference to the arguments of 'philosophers' concerning the origin of the idea of space appears in the concluding paragraph of section 1. Before it, Hume has stated the principle of the correspondence between feeling and thought: the 'general proposition', as he calls it,

That all our simple ideas in their first appearance are deriv'd from simple impressions, which are correspondent to them, and which they exactly represent. (*T*/4)

Ideas are occasioned by impressions and are true representations of them. A man born deaf or blind does not have any idea of sound or colour, nor can we have 'a just idea of the taste of a pine-apple, without having actually tasted it' (*T*/5).

After this, one finds something strange – a 'contradictory phenomenon'. It is the notorious case of a missing shade of blue in an otherwise continuous gradation of the same colour, and the admission that the imagination of someone well acquainted with all colours except that particular shade of blue would be able to fill the gap, that is, to form the idea of the missing shade, even in absence of the corresponding impression.

The case of the missing shade of blue is, it is tempting to say, a thought-experiment consisting in the construction of a counter-example to the principle which has just been introduced. This puzzling page is among the most discussed of Hume's whole philosophical writings. How is it that Hume produces this exception to his most basic and fundamental principle, and does so just after having introduced the principle itself? In addition Hume's 'thought-experiment', quite apart from the vexed issue of its imaginative performability, is apt to be regarded as extremely far-fetched.[5] It may be useful to concentrate for a moment on some of these problems.

For a start, if one takes into account the frame of reference of Hume's discussion, the hypothesis of the missing shade of blue turns out, in fact, not to be particularly far-fetched. It is, presumably, a Cartesian reminiscence – an analogy may be noted with Descartes' rule 14:

si quis à nativitate caecus sit, non sperandum est ullis unquam argumentis nos effecturos ut veras percipiat colorum ideas, quales nos habemus à sensibus

[5] See B. Rollin, 'Hume's blue patch', *Journal of the History of Ideas* 33, 1971, pp. 119–28. R. Fogelin, *Philosophical Interpretations*, Oxford 1992, chapter 4, 'Hume and the missing shade of blue', pp. 70–80. J. Passmore, *Hume's Intentions*, London 1980, 3rd edition, p. 93: 'insensitivity to consistency'; H. A. Pritchard, *Knowledge and Perception*, Oxford 1950, p. 177: 'effrontery'; Laird, *Hume's Philosophy*, p. 36: 'candour' (Laird also reports that there have been attempts to verify Hume's description of the phenomenon).

haustas; sed si quis primarios colores viderit quidem aliquando, intermedios autem et mixtos nunquam, fieri potest ut illorum etiam, quos non vidit, imagines ex aliorum similitudine per deductionem quamdam effingat.[6]

So, in introducing it, Hume is, in a way, answering a possible, probably fairly obvious and on the whole not very strong, objection against his proposition – the fact that the case of the missing shade of blue is very 'particular and singular' means that the principle may not be applicable in extreme cases, but does not affect its suitability for ordinary, common situations; and he insinuates that, however weak, such an exception is the strongest possible argument against his position. But this cannot be the whole story of the missing shade of blue. Anyone familiar with the *Treatise* will be aware of the frequency of the peculiar procedure of introducing counter-examples in the same breath as the general rule they violate. So the case of the missing shade of blue is not an isolated anomaly, but an instance of a recurrent rhetorical strategy, and as such requires careful consideration.

On the very first page of the *Treatise*, Hume says:

I believe it will not be very necessary to employ many words in explaining this distinction. Every one of himself will readily perceive the difference betwixt feeling and thinking. (*T*/1–2)

This is the first comment on the distinction – just introduced – between impressions and ideas: a distinction so obvious that it is not 'very necessary' to spend many words on it. And yet the passage carries on:

The common degrees of these are easily distinguished; tho' it is not impossible but in particular instances they may very nearly approach to each other. Thus in sleep, in a fever, in madness, or in any very violent emotion of the soul, our ideas may approach to our impressions: As on the other hand it sometimes happens, that our impressions are so faint and low, that we cannot distinguish them from our ideas. But notwithstanding this near resemblance in a few instances, they are in general so very different, that no-one can make a scruple to rank them under distinct heads, and assign to each a peculiar name to mark the difference. (*T*/2)

[6] *Regulae ad Directionem Ingenii*, Regula XIV [in the English translation of E. S. Haldane and G. R. T. Ross, *The Philosophical Works of Descartes*, vol. 1, Cambridge 1911, *Rules for the Direction of the Mind*, pp. 54–5: 'For example if a man has been blind from his birth it is not to be expected that we shall be able by any train of reasoning to make him perceive the true ideas of the colours which we have derived from our senses. But if a man has indeed once perceived the primary colours, though he has never seen the intermediate or mixed tints, it is possible for him to construct the images of those which he has not seen from their likeness to the others, by a sort of deduction'].

This sceptical commonplace seems to reverse the situation, obscuring the distinction; in fact it is once again, as in the case of the missing shade of blue, simply limiting the validity of the distinction between feeling and thinking to ordinary, everyday, situations.

There is one other case that I find revealing. It is in Book 2, where, after having established in its most delicate modulations the mechanics of the relationship linking pride with happiness and humility with misery, Hume concludes, in a moralising tone, with the reflection that

> the persons, who are proudest, and who in the eye of the world have most reason for their pride, are not always the happiest; nor the most humble always the most miserable, as may at first sight be imagin'd from this system. An evil may be real, tho' its cause has no relation to us: It may be real, without being peculiar: It may be real, without shewing itself to others: It may be real, without being constant: And it may be real, without falling under the general rules. Such evils as these will not fail to render us miserable, tho' they have little tendency to diminish pride: And perhaps the most real and the most solid evils of this life will be found of this nature. (*T*/294)

It is a reflection, Richetti comments, 'which warns readers not to suppose these theories cancel human irregularity'.[7] After all, 'the essence of the mind [is] equally unknown to us with that of external bodies', Hume writes in the Introduction to the *Treatise*; and all we can do is to 'endeavour to render all our principles as universal as possible' (*T*/xvii).[8] The first principle – as well as the other principles of the science of human nature – strikes a balance between human irregularity and the need for generality in the study of human nature. But what then does 'principle' mean in the pages of the *Treatise*?

Let us examine what terms are used in section i to refer to the principle of the correspondence between impressions and ideas. First of all, it is called a 'general decision' about the 'order and arrangement [of] our objects' (*T*/2–3) and a 'rule' that 'holds without any exception' between simple impressions and ideas (*T*/3). Later in the section it is a 'general proposition' of which 'we shall here content ourselves' since 'the *full* examination of this question is the subject of the present treatise' (*T*/4), and a 'general maxim' which it is not worth altering on the basis of the exception of the missing shade of blue (*T*/6). Finally it is 'the first principle I establish in the science of human nature':

[7] J. J. Richetti, *Philosophical Writing*, Cambridge (Mass.) 1983, p. 233.

[8] Cf. T/272: with the science of human nature 'we might hope to establish a system or set of opinions, which if not true (for that, perhaps, is too much to be hop'd for) might at least be satisfactory to the human mind, and might stand the test of the most critical examination'.

our simple impressions are prior to their correspondent ideas, and . . . the exceptions are very rare. (*T*/7)

This variety may appear confusing; in fact it provides, I suggest, the key to the status of the first principle. The first principle is called a 'maxim': a term that one frequently finds associated with 'proverb' in writings of the period. In Johnson's *Dictionary*, on the other hand, a 'maxim' is defined as an 'axiom', that is 'something evident at first sight', 'an established principle to be granted without new proof'. A moralistic colour is also apparent in 'rule', which means 'canon; praecept by which the thoughts or actions are directed', or 'regularity; propriety of behaviour', and in 'principle' itself, which is, in turn, a 'fundamental truth', an 'original postulate', a 'first position from which others are deduced', with the twofold active connotation of 'original cause', on the one side, and of 'ground of action; motive', on the other.[9] If one considers again the sententious tone of section 1 – with the appearance of candour and the naïveté of its 'egotisms', of its addresses to the reader, of its at once laborious and popular counter-examples – the hypothesis that the first principle is presented as a precept of human wisdom, something that any reader can easily recognise, accept, and use, becomes even more convincing. If this is the case, it is evident that the first principle is not meant to be exceptionless.

So, exceptions to the first principle are not the end of the world. This, however, does not mean that one can light-heartedly decide that the idea of space is an exception. For exceptions must be uncommon, irregular cases – there is a right kind of exception, of which the missing shade of blue is a perfect exemplar: artificial and peculiar at the same time. This is certainly not the case with the idea of space.[10] Besides, the mention of the idea of space as an illustration of the first principle in section 1 makes it difficult – very costly in terms of the plausibility of the resulting interpretation – to see it as an exception. The present difficulty brings us back to the treatment of the origin of the idea of space in Part 2, section 3, where, I believe, we are to look for an answer. Before turning to it, however, I intend to linger for a moment on the

[9] S. Johnson, *A Dictionary of the English Language*, London 1755, s.v. 'Maxim', 'Axiom', 'Rule', and 'Principle'. On 'rules' and 'general rules' in the *Treatise* see below, chapter 3, 'Hume's a priori between Kant and Husserl', and fn. 37.

[10] See D. Garrett, *Cognition and Commitment in Hume's Philosophy*, Oxford 1997, pp. 50ff., for a masterly recent analysis of the missing shade of blue. Garrett's conclusion is similar to mine: in his opinion, the missing shade of blue 'has little effect on Hume's philosophical uses of the principle', 'because of its special explanations and near-miss character'. He also thinks that the ideas of space, time, and existence would be 'far more serious breaches of the Copy principle' (p. 52).

pages where the first principle is introduced, in order to brood on some curious features of the philosophical language in which they are written. Doing so will provide an important clue concerning the main problem with the origin of the idea of space.

DIVAGATION ON HUME'S CHOICE OF WORDS

I find it interesting that in the first pages of the *Treatise* Hume's language is at once very precise and very loose. The looseness of the language of the *Treatise* is notorious, and yet Hume's choice of words is, from a certain point of view, far from careless. It is worth considering in some detail the conclusion of section 1, that is, the paragraph which follows the treatment of the missing shade of blue. Hume explains that

> the present question concerning the precedency of our impressions or ideas, is the same with what has made so much noise in other terms, when it has been disputed whether there be any *innate ideas*, or whether all ideas be derived from sensation and reflexion. (*T*/7)

On the one hand, by bringing his approach to bear on the classic metaphysical question of innate ideas, Hume seems to employ a common rhetorical device to legitimate a new approach, namely that of setting it against precedents, with the intention of making clear at the same time its originality and its continuity with tradition. On the other hand, such a reference obviously implies an allusion to, and a criticism of, Locke's treatment of the problem of innate ideas. (This was evidently quite a bold move, since Book 1 of Locke's *Essay* was widely considered as the last word on the matter.[11]) In fact, with the distinction between impressions and ideas, Hume has written a few pages earlier:

> I rather restore the word, idea, to its original sense, from which Mr Locke had perverted it, in making it stand for all our perceptions. (*T*/2 n.)

The significance of the relation between impressions and ideas is shown by pointing out that it makes it possible to distinguish two kinds of mental contents or perceptions: the ones that 'arise in the soul originally, from unknown causes' (*T*/7) – i.e. the impressions, which, from this point of view, may be said to be innate – and those that can be

[11] For example, Edmund Law, in his edition of William King's *De origine mali*, cuts the author's discussion of innate ideas, and explains: 'I omit the rest of our Author's long Note, since it contains only some Arguments for *innate Ideas*; which Hypothesis is now almost universally exploded'. Law proceeds then to explain King's 'different Signification of the Word *Idea*' and to compare it closely with Locke's (*An Essay on the Origin of Evil*. By Dr. William King, late Lord Archbishop of Dublin. Translated from the Latin, with Large Notes; . . . London 1731, pp. 6ff.).

referred to something more original acting as a cause – i.e. the ideas, which are said to derive from correspondent impressions. So, the restoration of the word 'idea' to its original sense and the distinction between ideas and impressions have the effect of radically reshaping the question of innate ideas, and of questioning again Locke's answer to it: its presuppositions are subverted. It is at once the acknowledgment of the importance of Locke's position, used as the polemical target and thus the point of reference for Hume's own approach, and the dissolution of the whole problem through a change in the metaphysical terminology:

> I hope this clear stating of the question [of innate ideas] will remove all disputes concerning it and will render this principle of more use in our reasonings, than it seems hitherto to have been. (*T*/7)

There is also, here, a hint at the extent of the future role of the principle: up to this point Hume has simply introduced the principle and faced it with a question – that of innate ideas – which can be considered as a test of its power and utility. Later on he will show in what way the application of the first principle is going to reshape the treatment of metaphysical questions; and the first principle is presented as a change in the basic language in which they are expressed.

So, the distinction between impressions and ideas is meant to introduce a crucial change in metaphysical terminology, one which shifts metaphysical problems dramatically. Some perceptions, the impressions, are original, while others, the ideas, are copies. Impressions are not 'derived from sensation and reflexion', but arise 'originally' and 'from unknown causes': but then, for all we know, the mind is not an empty cabinet, it may well be endowed with some active power or quality.

Hume has clearly stated that 'impressions' and 'ideas' are technical terms by saying that they are used

> in a sense different from what is usual, and I hope this liberty will be allowed to me. Perhaps I rather restore the word, idea, to its original sense, from which Mr. *Locke* had perverted it, in making it stand for all our perceptions. By the term of impression I would not be understood to express the manner, in which our lively perceptions are produced in the soul, but merely the perceptions themselves; for which there is no particular name either in the *English* or any other language, that I know of. (*T*/2 n.)

Hume's attention to terminology is also suggested by the corresponding remark in the *Enquiry*, section 2, which reproduces, without any relevant change, the approach and the argumentative structure of section I of the *Treatise*. After the discussion of the missing shade of blue (which is actually taken almost word for word from the *Treatise*, with some alterations in punctuation and spelling), one finds in the concluding footnote the corresponding reference to the problem of innate ideas; and this time Locke's name appears openly:

> It is probable that no more was meant by those, who denied innate ideas, than that all our ideas were copies of our impressions; though it must be confessed, that the terms, which they employed, were not chosen with such caution, nor so exactly defined, as to prevent all mistakes about their doctrine. For what is meant by *innate*? . . . Again, the word *idea*, seems to be commonly taken in a very loose sense, by Locke and others; as standing for any of our perceptions, our sensations and passions, as well as thoughts . . . admitting these terms, *impressions* and *ideas*, in the sense above explained, and understanding by *innate*, what is original or copied from no precendent perception, then may we assert that all our impressions are innate, and our ideas are not innate. To be ingenuous, I must own it to be my opinion, that *Locke* was betrayed into this question by the schoolmen, who, making use of undefined terms, draw out their disputes to a tedious length, without ever touching the point in question. A like ambiguity and circumlocution seems to run through that philosopher's readings on this as well as most other subjects. (*E*/17 n.)

Hume's attention is here so focussed on terminology that he considers loose language a relevant shortcoming of Locke's philosophical writings.

On the other hand, in the very same section of the *Treatise* it is possible to find particularly good examples of Hume's own loose language. For instance, the terms employed to describe and compare impressions and ideas are: 'force', 'violence', 'liveliness', 'vivacity', and 'strength', and 'faintness', 'weakness', and 'languid'. It is true that this vagueness is balanced by the brief discussion of innate ideas: this makes it clear that impressions are different from ideas because they are original ('innate') while ideas are copies, so that the difference in force and liveliness may be taken as a more empirical criterion for the distinction, a kind of rule of thumb. But, more seriously, the same terminology appears in all its vagueness also (to limit myself to Part I) when Hume talks about the difference between ideas of the memory and ideas of the imagination, and about the mechanism at the origin of abstract ideas.[12] Selby-Bigge wrote that

[12] See Stroud, *Hume*, pp. 28ff., for a typical criticism of such terms, considered as inadequate to express the ground for the distinction between feeling and thinking. Passmore, *Intentions*, pp. 93–4, suggests that the distinction between impressions and ideas and the criteria for it are presented with such 'casualness and brevity' because 'Hume was introducing a new terminology

> though [Hume's] language is often perversely loose, yet it is not always the expression of loose thinking.[13]

In the case of the set of terms we are considering, Hume seems to know what he is doing, and why. A full explanation of his loose language is to be found in the addition (*T*/628–9), where, in commenting on his own way of discussing belief, he says:

> This operation of the mind, which forms the belief of any matter of fact, seems hitherto to have been one of the greatest mysteries of philosophy; tho' no one has so much as suspected, that there was any difficulty in explaining it. For my part I must own, that I find considerable difficulty in the case; and that even when I think I understand the subject perfectly, I am at a loss for terms to express my meaning . . . I scarce find any word that fully answers the case, but am oblig'd to have recourse to every one's feeling, in order to give him a perfect notion of this operation of the mind. An idea assented to *feels* different from a fictitious idea, that the fancy alone presents to us: And this different feeling I endeavour to explain by calling it a superior *force*, or *vivacity*, or *solidity*, or *firmness*, or *steadiness*. This variety of terms, which may seem so unphilosophical, is intended only to express the act of the mind, which renders realities more present to us than fictions, causes them to weigh more in the thought, and gives them a superior influence on the passions and imagination. Provided we agree about the thing, 'tis needless to dispute about the terms . . . I confess, that 'tis impossible to explain perfectly this feeling or manner of conception. We may make use of words, that express something near it. But its true and proper name is *belief*, which is a term that every one sufficiently understands in common life. And in philosophy we can go no farther, than assert, that it is something felt by the mind which distinguishes the ideas of the judgment from the fictions of the imagination. It gives them more force and influence; makes them appear of greater importance; infixes them in the mind; and renders them the governing principles of all our actions.

One may well remark on this passage borrowing Hume's own words: his language 'only touches the soul, if I may be allow'd so to speak' (*T*/20).[14]

The cases I have listed of loose language in the *Treatise* have much in

rather than a new doctrine . . . all this, except the terminology, was so familiar that Hume could mention, merely as a curiosity, that there were sometimes ideas which did not seem to copy any impression.' But see R. J. Butler, 'Hume's impressions', in G. Vesey (ed.), *Impressions of Empiricism*, London 1976, pp. 122–36.

[13] In the Preface to his edition of the *Treatise*.

[14] Kemp Smith, *Philosophy*, p. 110, considers the language of section 1 as the mark of the first, though disguised, appearance of belief in the *Treatise*. See also the stimulating reading proposed by D. Livingston, *Hume's Philosophy of Common Life*, Chicago 1984, pp. 57ff. (on p. 58, concerning *T*/628–29, Livingston writes: 'we may see in this variety what Husserl saw: a pioneering attempt at phenomenological description').

common – in fact, they are all recurrences of a true family of loose, evocative, or allusive terms, terms which 'touch the soul'. These terms are sometimes introduced or accompanied by open expressions of regret or apologies for their inefficacy, insufficiency, and so on; and, even when no such apology appears, their use is invariably excused with the fact that what is being said anyway is well known and evident. Still, no matter how open the recognition of their limitations, they never fail to be there when the circumstances require them: that is, when the subject is something at once very familiar and difficult to express – something at once intimate and ineffable, something which has to do not with ideas, but with, for instance, the differences between how different kinds of ideas are felt. I suggest that, in the *Treatise*, the language becomes loose when mental contents are being described not as such, but in terms of qualities and acts of the mind. The action of the mind is described as:

> That certain *je-ne-scai-quoi*, of which it is impossible to give any definition or description, but which every one sufficiently understands. (*T*/106)

This 'certain *je-ne-scai-quoi*' is not the most frequently used member of Hume's family of loose terms; but it looks by no means out of place among 'force', 'vivacity', 'this different feeling', 'this feeling or manner of conception', and so on. That to describe an action of the mind, a 'certain *je-ne-scai-quoi*', one may end up resorting to some loose language, is not surprising. Something of this kind – a certain tendency to use these loose, evocative, and allusive expressions – also appears in connection with the origin of the idea of space. It is now time to turn to the idea of space itself, and to Hume's account of its origin.

A PHILOSOPHER'S EYE

It is now worth reading how the treatment of the origin of the idea of space begins:

> Let us apply this principle [that every idea, with which the imagination is furnish'd, first makes its appearance in a correspondent impression], in order to discover further the nature of our ideas of space and time. Upon opening my eyes, and turning them to the surrounding objects, I perceive many visible bodies; and upon shutting them again, and considering the distance betwixt these bodies, I acquire the idea of extension. As every idea is deriv'd from some impression, which is exactly similar to it, the impression similar to this idea of extension, must either be some sensations deriv'd from the sight, or some

internal impression arising from these sensations. Our internal impressions are our passions, emotions, desires and aversions; none of which, I believe, will ever be asserted to be the model, from which the idea of space is deriv'd. There remains therefore nothing but the senses, which can convey to us this original impression. (*T*/ 33)

Hume says that he is applying his first principle to find out the origin of the idea of space, and thus to substantiate the claim that such an idea is 'conveyed by our senses'. In the first place, the idea of space or extension is said to arise after shutting our eyes and 'considering' – examining, pondering – the distance between the impressions of the bodies; the idea of space appears as a consequence of sense-experience. Hume is talking the language of abstraction. His 'considering' is curious and revealing. It is a sign that the origin of the idea of space is more complex a subject than the passage above may suggest.[15]

In fact, the experience presented here is anything but simple. In a way, it is precisely its everyday, familiar aspect which makes it too complex for the purpose: it is an actual, if elementary, example of sense-perception, taken unanalysed. For example, what are the 'visible bodies' and 'the distance betwixt these bodies' perceived by the eye? The answer in terms of Hume's science of human nature is that they, bodies and distances alike, are nothing but coloured shapes on the retina. This means that their identification as bodies and distances requires an act of the judgement; and, in general:

Those, who are acquainted with the metaphysical part of optics . . . know how we transfer the judgments and conclusions of the understanding to the senses. (*T*/374–5)

This point may be made more clear by reference to Hume's treatment of the visual field as a painted surface in connection with an analogous problem, that of the assessment of the distance of bodies from us:

'Tis commonly allow'd by philosophers, that all bodies, which discover themselves to the eye, appear as if painted on a plain surface, and that their different degrees of remoteness from ourselves are discover'd more by reason than by the senses. When I hold up my hand before me, and spread my fingers, they

[15] I am grateful to John Yolton for asking me interesting questions about the appearance of 'considering' in this passage. Michael Ayers reminded me that it is usual to find the term in discussions of abstraction (and suggested the comparison with Locke's 'partial consideration'). For 'to consider' Johnson's *Dictionary* gives 'to think upon with care; to ponder; to examine; to sift; to study' as primary meanings; other meanings are: 'to take into view; not to omit in the examination', 'to have regard to; to respect . . .', 'to think maturely; not to judge hastily or rashly', 'to deliberate; to work in the mind', 'to doubt; to hesitate'.

are separated as perfectly by the blue colour of the firmament, as they cou'd be by any visible object, which I cou'd place betwixt them. (*T*/56)

From this passage it appears that recognition of bodies and the distances between them – my fingers and the blue sky by which they are separated – shows a significant similarity to assessment of the distance of bodies from the perceiving eye. The eye, if unassisted by judgement, would only see coloured patches – say, a splash of a pinkish colour and light blue. Judgement identifies the blue patch as sky, and the pinkish patch as my hand with fingers spread against the sky; this involves an assessment of the different distances of the sky and of my fingers from my eye, that is, from myself; and this is the basis for calling the blue patches not only 'sky', but also, in a different context, what separates my fingers, the 'distance between them'.[16]

So, in the everyday example, where the idea of space is derived from the distances between the bodies surrounding us, the problem is that sense-perception and the act of the judgement are not distinguished, but are simply described as they occur in ordinary experience. This illustration of the origin of the idea of space is not yet sufficiently abstract, since the presence of an act of judgement interferes with the observation: the mention of the distances between the bodies, in particular, is a clear sign of this interference. In addition, in this first instance the empirical origin of the idea of space is shown unequivocally, but – not surprisingly, given the complexity of the case – the way in which that idea is derived from sense-data is still far from clear: what impression is at the origin of the idea of space? With this treatment, in short, the discussion does not go much beyond a general empirical attitude to the problem, as it was expressed in section 1 – except in allowing a quick dismissal of the impressions of reflection ('passions, emotions, desires and aversions') as implausible candidates for the origin of the idea of space. This last is a point which deserves a moment more of attention.

The exclusion of passions as a possible origin says something interesting about the idea of space: it means that its case is different from that of other ideas whose origin is for some reason difficult to find. It is different, in particular, from the idea of necessary connection, which has its origin in an impression in a perfectly normal way, except that the impression is, as it were, not where one would look for it. The origin of

[16] See above, chapter 1, 'Units of sight', for comments on another aspect of this, and below, chapter 4, 'Simplicity and distance', for a fuller treatment of bodies and distance in connection with the discussion of the idea of an empty space.

the idea of necessary connection is an impression of reflection – the 'propensity, which custom produces, to pass from an object to the idea of its usual attendant' (*T*/165) – which is commonly misunderstood and projected outside the mind.[17] In the case of the idea of space, one is told that no such impression is to be found: the origin of the idea of space must be found in the senses.[18] But how?

After what I have called the everyday example, the argument goes on as follows:

> Now what impression do our senses here convey to us? This is the principal question, and decides without appeal concerning the nature of the idea. The table before me is alone sufficient by its view to give me the idea of extension. This idea, then, is borrow'd from, and represents some impression, which this moment appears to the senses. But my senses convey to me only the impressions of colour'd points, dispos'd in a certain manner. If the eye is sensible of any thing farther, I desire it may be pointed out to me. But if it be impossible to shew any thing farther, we may conclude with certainty, that the idea of extension is nothing but a copy of these colour'd points, and of the manner of their appearance. (*T*/33–4)

Here, what is being scrutinised is a complete, but much simplified, example of a sense-experience immediately felt as spatial. Obviously its simplicity is meant to make the study of the matter easier. The features of the example of the table are very different from those of the surrounding bodies. The kind of experience involved here is highly abstract; it is the creation of an artificial model of the first occurrence of the idea of space to the mind. The example is chosen in such a way that the result of the decomposition of the visual field into its elementary contents is not ordinary objects and their relative positions, distant bodies and their distances. In the first example, one of the problems was that bodies and distances were distracting and misleading because too familiar. This time the objects spatially related are, in fact, elementary

[17] On this point see the stimulating views of R. P. Wolff, 'Hume's theory of mental activity', in V. C. Chappell, *Hume*, London 1966, pp. 99–128. Wolff considers the treatment of the idea of necessary connection as 'the precise point at which the framework of [Hume's] system distorts his exposition and forces him to cramp his thoughts into ill-suited categories' (p. 100). More specifically, when talking about the impression of reflection at the origin of the idea of necessary connection 'we can observe Hume shifting towards explanation in terms of mental activities, while still tied to the language of mental contents' (p. 112). See also below, chapter 3, especially the paragraph on 'A passion for truth'.

[18] The same is also explicitly stated about the idea of time: 'Five notes play'd on a flute give us the impression and idea of time; tho' time be not a sixth impression, which presents itself to the hearing or any other of the senses. Nor is it a sixth impression, which the mind by reflection finds in itself' (*T*/36).

perceptual units, elements of a coloured shape on the retina. So there is an object easily identifiable as such – the surface of the table – only at the beginning of the analysis; at the end the table has been eliminated and one is left with unfamiliar, philosophical entities – coloured and tangible points. In this case, no acts of judgement are implied in the reaction of the mind to the sense-experience. While looking at the table to inquire into the origin of the idea of space, Hume brackets all that is in his mind already concerning his table, tables in general, colours, the experience of objects, and the usual interpretation of the coloured shapes on the retina in terms of external reality, that is, as bodies and distances between them. He forces his own mind to react according to its bare natural principles and tendencies at the most elementary level. Such an unprejudiced observation of experiential data is anything but commonsensical, spontaneous, and easy. The description of the origin of the idea of space from the sight of the table is an instance of what Hume thinks of as 'experimental method of reasoning'; it is an 'experiment' in the science of human nature. And the first principle is the tool that guides this experiment and makes it effective.[19]

The result of this analysis is also very abstract. Being extended is being an object of sight and touch, perceived through such specific channels:

> The idea of space is convey'd to the mind by two senses, the sight and touch; nor does any thing ever appear extended, that is not either visible or tangible. (*T*/38; cf. *T*/235–6)

This is the common feature, the resemblance at the basis of the idea of space. The next step in the experimental reconstruction of the origin of the idea of space is to show how, from several different experiences of visual, that is spatial, perceptions, we form the abstract idea of space based

[19] For a different view of Hume's experiments see Noxon, *Development*, p. 117: experiment in Hume's sense 'is a synonym for "experience", and conforms to the usage of Thomas Hobbes and John Locke rather than to that of Robert Hooke or of Robert Boyle. Observant, reflective, introspective, Hume was no experimenter in the modern sense given the term by the Royal Society. In an age when experimental work made scientific reputations, and even amateurs delighted in repeating the experiments reported by professionals, Hume remained a philosophical thinker, a stranger to the "careful and exact experiments" which he extolled . . . One need only to compare the suggested image of Hume at his desk poking his finger in his eye with the picture . . . of Robert Boyle at work with his cylinders of mercury and airpump, his "exact pair of Ballances" and aeolipile, to grasp the magnitude of the difference between the two men's conception of a scientific experiment' (p. 118; the reference is to *T*/210–11). Further, according to Noxon, all Hume's 'experiments' in Part 2, section 2 of the Book on passions are, in fact, stories, and 'The point is not that Hume's stories are improbable. The point is that stories are not experiments' (p. 119).

merely on that disposition of points, or manner of appearance, in which they [the extended objects, or compositions of coloured points] agree. (*T*/34)

Such an abstract idea is then enlarged to include the impressions of touch as well as those of sight, 'upon account of their resemblance'. (*T*/34)

What I find particularly interesting in this account of the idea of space as an abstract idea is the reference to the impressions of touch. However brief, this reference means that 'manner of appearance' cannot be simply referred to the physiology of vision, or be directly identified with what happens in the eye. Only the case of sight is treated extensively, and in what sense the impressions of touch can be 'found to be similar to those of sight in the disposition of their parts' is not spelled out. But it is clear that the resemblance at the origin of the abstract idea of space cuts across the modes of operation of two senses, and must be more complex, more original, and, as it were, located at a more basic level than either of them. At this point, in order to approach the source of this resemblance and of the idea of space, another (apparent) divagation on the language of the *Treatise* will not be out of place.

SECOND DIVAGATION ON HUME'S CHOICE OF WORDS

'Dispos'd in a certain manner', 'this particular manner of appearance', it is impossible not to notice, are far from the crisp phrases one might expect as the outcome of the application of the first principle to a straightforward case. They are allusive, evocative expressions: yet again, as when he describes the difference between impressions and ideas, Hume's language is loose. The examination of the case of the table in terms of the first principle leaves a residue, something which cannot be resolved. The existence of this residue suggests the right questions to address to these pages of the *Treatise*: how far can an inquiry guided by the first principle go in accounting for the origin of the idea of space? And why is the idea of space so difficult to treat?

The idea of time, Hume says,

> is not deriv'd from a particular impression mix'd up with others, and plainly distinguishable from them; but arises altogether from the manner, in which impressions appear to the mind, without making one of the number: (*T*/36)

and

> As 'tis from the disposition of visible and tangible objects we receive the idea of space, so from the succession of ideas and impressions we form the idea of

time, nor is it possible for time alone ever to make its appearance, or be taken notice of by the mind. (*T*/35)

The ideas of space and time arise from the manners of appearance of impressions, and are the ideas of the manners or orders of existence of original perceptual objects – they are 'no separate or distinct ideas', but only appear as features of complex impressions, and are both occasioned and constituted by the complex impressions. As such, they seem entirely to dissolve when one uses the first principle to analyse those complex impressions. I propose that we regard this arrangement of simple visual and tactile impressions, since it only characterises the totality of a complex impression, as the expression of an original contribution of the mind to sense-experience. From this point of view it is interesting to note the similarity between Hume's description of the ideas of space and time and his description of the self:

self or person is not any one impression, but that to which our several impressions and ideas are suppos'd to have a reference. (*T*/251)

But then, I suggest, in the inquiry into the origin of the idea of space (and time), the use of the first principle leads to the discovery of two facts: first, the simple elements, the coloured and tangible points into which the table is dissolved; second, the limits of application of the principle itself. According to this reading, the ideas of space and time are ideas of a particular kind, fundamentally different from other ideas: they are not mental contents in the ordinary sense, because they are somehow more directly and specifically related to the self. In this sense, the farthest the first principle can go concerning matters like the origin of the idea of space is to show that the idea of space is, as it were, a point of singularity in our experience.

And, if this is so, one must acknowledge that the first principle has done its job well. It has directed the experiment of the table. It has made it possible to discover and to bracket the mental contents – the elementary perceptions of sight and touch – involved in the origin of the idea of space, and thus to single out the act of the mind as a residue. The first principle cannot properly account – nor is it supposed to – for either the structure, or even the 'powers and qualities' (as Hume calls them) of the mind. For,

the essence of the mind being equally unknown to us with that of external bodies, it must be equally impossible to form any notion of its powers and qualities otherwise than from careful and exact experiments, and

> any hypothesis, that pretends to discover the ultimate original qualities of human nature, ought at first to be rejected as presumptuous and chimerical. (*T*/XVII)

It is now time to turn to the self, and to the traces of its powers and qualities as they appear in the *Treatise.*

MANNERS OF APPEARANCE AND MANNERS OF CONCEPTION

I have suggested that the idea of space, in so far as it is constituted not only by perceptual objects, but also by their manner of appearance, is the expression of a power or quality of the mind; and that, as such, it may be considered as the original contribution of the mind to sense-perception. This is why the ideas of space and time are, as I called them earlier, very special. But in the *Treatise* there is something else sharing their special position, though dissimilar from them in other respects. As impressions are organised in spatial and temporal manners of appearance, so ideas are characterised by the way they are felt – the manner of their conception. This manner of conception is belief.

Belief is described with the typical terminology used in talking of powers and qualities of the mind. It is defined as 'somewhat more than a simple idea. 'Tis a particular manner of forming an idea' (*T*/97), 'a more vivid and intense conception of an idea, proceeding from its relation to a present impression' (*T*/103), 'some sensation or peculiar manner of conception' (*T*/184):

> As 'tis certain there is a great difference betwixt the simple conception of the existence of an object, and the belief of it, and as this difference lies not in the parts or composition of the idea, which we conceive; it follows, that it must lie in the *manner*, in which we conceive it. (*T*/94–5)

It is now clear that the problems of the treatment of the idea of space as a manner of appearance are not the problems of an isolated, marginal case; they are general, and crucially important. At the core of the metaphysics of the *Treatise* there is, I suggest, something that cannot be further resolved or referred to anything else external to it – what I have called the original contribution of the self, whose presence in the text is signalled by evocative terms. This is because, however familiar, this contribution cannot be spelled out in terms of the first principle, but only detected as a residue. If this is the case, one more thing becomes clear – something about the way the self is treated in Part 4 of the *Treatise*.

Hume's treatment of the self is one of the most difficult and most discussed aspects of his thought. As so often, there is the appearance of a flat contradiction. In Book 1 of the *Treatise*, in the sections on the immateriality of the soul and on personal identity, there is the well-known statement that, since no distinct impression can be found at the origin of the idea of the self, the self is nothing but 'a connected heap of perceptions' (*T*/207); while, in Book 2, Hume scatters a number of remarks to the effect that our perception of our own self is always present to us, and is particularly strong and lively. Now, this does look like a contradiction.[20] But what I have suggested about manners of appearance and of conception may afford the key to Hume's position on this point.

The clue is, in my opinion, a passage from the section on personal identity, where one is told that

> we must distinguish betwixt personal identity, as it regards our thought or imagination, and as it regards our passions or the concern we take in ourselves. The first is our present subject. (*T*/253)

This seems to mean that the apparent contradiction between the treatment of self in Book 1 and Book 2 is, in fact, the deliberate expression of a metaphysical problem.[21]

This impression may be reinforced by considering some aspects of Hume's treatment of personal identity. It is important to remember that the polemical target of the section on personal identity is not the existence of a conception of the self, but rather the theory of the identity and simplicity of a substantial soul. Hume is writing against

20 Observed, for instance, by T. E. Jessop, 'The misunderstood Hume', in W. B. Todd (ed.), *Hume and the Enlightenment*, Edinburgh 1974; Noxon, *Development*, p. 19: 'One would expect . . . the constructive side of the argument to yield a concept of mind or self which Hume could use in his study of the passions . . . But his study "Of personal identity" is dominated by the analytical, critical mood, and the result appears to be purely sceptical, purely negative.' For a different point of view, which is closer to the reading I am proposing here, see especially N. Capaldi, 'The historical and philosophical significance of Hume's theory of the self', in A. J. Holland (ed.), *Philosophy, Its History and Historiography*, Dordrecht 1985, pp. 271–85; and A. C. Baier, *A Progress of Sentiments*, Cambridge (Mass.) 1991, pp. 129ff.: 'there is . . . no contradiction between what Book I said about the self and what Book II says; rather, there is supplementation and completion' (p. 129); see also Richetti, *Writing*, pp. 218ff.

21 Passmore, *Intentions*, pp. 79ff. notes the crucial importance of *T*/253, but says that 'the fact remains that [Hume] never brings the two sorts of personal identity into relation with one another' (p. 79).

some philosophers, who . . . are certain, beyond the evidence of a demonstration, both of its perfect identity and simplicity.[22] (*T*/251)

So the scepticism of this section has a limited scope. The refutation is guided by the first principle:

From what impression cou'd this idea be deriv'd? . . . It must be some one impression, that gives rise to every real idea. (*T*/251)

But

when I enter most intimately into what I call *myself*, I always stumble on some particular perception or other . . . I never catch *myself* at any time without a perception, and can never observe anything but the perception. (*T*/252)

Thus the use of the first principle shows the untenability of the theory of the self as an identical simple substance.

After this refutation, the argumentative strategy comes as a surprise.[23] The foundation of the self is not introspective. Of course, an inquiring mind is presupposed all along.[24] But it only makes its appearance obliquely: it is not a metaphysically charged 'I', but a neutral, less committed 'we' – the author of the *Treatise* and its readers. It does not appear as the metaphysical subject, but as our mind, which naturally performs the reading and (even more importantly) the misreading of signs, and thus conceives the idea of identity. The starting-point of the discussion is the identity we attribute to uninterrupted and invariable objects (*T*/253). But, since we tend to confound identity with relation, we attribute identity to interrupted and variable (but related) objects as well (*T*/254). This tendency is so powerful, that 'our mistake' is usually also accompanied by the fiction of

something unknown and mysterious, connecting the parts, beside their relation; and this I take to be the case with regard to the identity we ascribe to plants and vegetables. (*T*/254–5)

22 As Capaldi observes in his 'Significance', p. 272. See also Kemp Smith, *Philosophy*, pp. 96ff. and 500.

23 This paragraph owes much to a suggestion of John Yolton, who pointed out to me the significance of the similarity between identity in plants and animals and identity of persons in this section of the *Treatise*.

24 Cf. Passmore, *Intentions*, who maintains that, in Hume's analysis of causal necessity, substance, belief in the existence of the external objects, and, generally, 'through the whole of Hume's psychological positivism' (p. 78), there is the assumption of 'a self over and above the order of our perceptions' (p. 77), as the subject 'which can be affected by the occurrence . . . of a C followed by an E', which can confuse external objects and impressions – 'if there is to be a confounding there must be someone who is confused' (p. 78).

There are many other circumstances which promote the confusion between identity and relation: for instance, the gradualness of the change. The most powerful of these circumstances is connected with the relation of cause and effect: the '*sympathy* of parts to their *common end*', which appears in plants and animals. It is in this connection, in a comparison with a plant, that a person (other than the narrating 'we') first appears in this analysis:

> tho' every one must allow, that in a very few years both vegetables and animals endure a *total* change, yet we still attribute identity to them, while their form, size, and substance are entirely alter'd. An oak, that grows from a small plant to a large tree, is still the same oak; tho' there be not one particle of matter, or figure of its parts the same. An infant becomes a man, and is sometimes fat, sometimes lean, without any change in his identity. (*T*/257)

In a similar way, the confusion of numerical and specific identity makes us call 'a noise, that is frequently interrupted and renew'd . . . the same noise' (*T*/258). Again a considerable change, expected because regarded as 'natural and essential' to the thing, 'by breaking less the continuity of the thought, has less influence in destroying the identity': so that

> as the nature of a river consists in the motion and change of parts; tho' in less than four and twenty hours these be totally alter'd; this hinders not the river from continuing the same during several ages. (*T*/258)

It is only after these lively pages full of classical, very literary examples, that the discussion focusses on personal identity:

> The identity, which we ascribe to the mind of man, is only a fictitious one, and of a like kind with that which we ascribe to vegetables and animal bodies. (*T*/259)

Hume is still discussing not his own self, but the personal identity of somebody else. His question at this point is:

> whether in pronouncing concerning the identity of a person, we observe some real bond among his perceptions, or only feel one among the ideas we form of them. (*T*/259)

The answer is again in the same tone:

> identity is nothing really belonging to these different perceptions, and uniting them together; but is merely a quality, which we attribute to them, because of the union of their ideas in the imagination, when we reflect upon them . . . The only question, therefore, which remains, is, by what relations this uninter-

rupted progress of our thought is produc'd, when we consider the successive existence of a mind or thinking person. And here 'tis evident we must confine ourselves to resemblance and causation. (*T*/260)

And finally – after an analysis of resemblance and the role of memory in the following terms: 'suppose we cou'd see clearly into the breast of another' (*T*/260) – one finds a short, absent-minded remark: 'The case is the same whether we consider ourselves or others' (*T*/261). It is evident that in this analysis a convincing and living self only appears under the disguise of the inconspicuous narrating 'we', the unnoticed self that impersonally and mercilessly analyses other identities, other selves, and its own natural errors – a self which never becomes the object of the analysis.

After all this, the self is intimated through three different images:

nothing but a bundle or collection of different perceptions, which succeed each other with an inconceivable rapidity, and are in a perpetual flux and movement: (*T*/252)

the mind is a kind of theatre, where several perceptions successively make their appearance; pass, re-pass, glide away, and mingle in an infinite variety of postures and situations: (*T*/253)

I cannot compare the soul more properly to any thing than to a republic or commonwealth, in which the several members are united by the reciprocal ties of government and subordination, and give rise to other persons, who propagate the same republic in the incessant changes of its parts . . . in this view our identity with regard to the passions serves to corroborate that with regard to the imagination. (*T*/261)

Much could be said about each of these three images – it would be interesting to observe in detail the shifts of emphasis from one to the other; or to trace them back to possible sources and to explore how frequently they appear in other texts of the period, in order to place Hume's treatment in its topical context (the commonwealth one, for example, appears in Bayle's *Dictionary*: 'the Inconstancy of Humane Opinions is so great, that one would think Man is a small Republick which often changes its Magistrates'[25]). Here it suffices to notice that, after the analysis in terms of the first principle, there comes not one, but three descriptions of the self, and that all of them are in metaphorical terms. Again, Hume is trying to find evocative words, words which can 'touch our soul'.

[25] 'Pyrrho', n. G, in the English edition, London 1710, p. 2623 (see Laird, *Hume's Philosophy*, p. 163).

From another point of view, of course, Hume can easily refer to the

> self, or that succession of related ideas and impressions, of which we have an intimate memory and consciousness: (*T*/277)

or even maintain that

> 'Tis evident, that the idea, or rather impression of ourselves is always intimately present to us, and that our consciousness gives us so lively a conception of our own person, that 'tis impossible to imagine, that any thing can in this particular go beyond it. (*T*/317)

But here it is passions that are under discussion; and this is why Hume can go so far as to absent-mindedly refer to 'the idea, or rather impression' of the self. The precise distinction between impression and idea does not matter in this connection: all that is left of it is 'a conception' that is better regarded as an impression rather than an idea, on account of its being presumably 'original', and in any event 'so lively'.[26]

On the other hand, in treating the understanding, the science of human nature has as its true objects the contents of the mind rather than its powers and qualities. It can only approach the self indirectly. No theory is possible, but only the dissolution of the mind into its elementary contents. What is left is a sort of side-effect of the study of mental contents: images and metaphors, alluding to the 'powers and qualities' of the mind. In proximity to the self, the language of the *Treatise* tends to become distorted or unfocussed in a characteristic way; and it is the self that is being alluded to in those passages of Book 1 where one is faced with this loose, evocative, allusive language. In the treatment of the origin of the ideas of space and time (as in that of belief) this typical language makes its appearance; and it does so because the treatment is entirely informed by the allusion to the 'powers and qualities' of the mind.

In the Addition to Book 1, Hume writes, concerning the theoretical treatment of the self, that

> I find myself involv'd in such a labyrinth, that, I must confess, I neither know how to correct my former opinions, nor to render them consistent, (*T*/633)

so that

> I must plead the privilege of a sceptic, and confess, that this difficulty is too hard for my understanding. (*T*/636)

[26] The same loose use of the term 'impression' also appears in *T*/156, about the idea of necessity. See also *T*/164ff., on necessary connection, and *T*/180–1 on the certainty of knowledge.

Expression of a 'metaphysical agony', or a piece of rhetoric?[27] However the matter stands, this page says once more that no theory can be produced concerning either the structure and essence, or the qualities and powers of the mind. We cannot even find names for the powers and qualities of the mind. The self arranges perceptions in certain orders or dispositions, and organises itself as a set of beliefs. And this is all that metaphysics can say.

Of course, there is more to this story. If one talks about passions, all this is left behind and there are no more contradictions, no more doubts. For then it is the right way of talking about the self that comes to the fore, which is not theoretical, but historical – as one may call it, following the sense in which Tristram Shandy calls Locke's *Essay* a history:

> I will tell you in three words what the book is. – It is a history. – A history! of who? what? where? when? Don't hurry yourself. – It is a history-book, Sir, (which may possibly recommend it to the world) of what passes in a man's own mind; and if you will say so much of the book, and no more, believe me, you will cut no contemptible figure in a metaphysic circle.[28]

A CAREFUL VAGUENESS

A very sensitive Hume interpreter has written that

> when Hume's words are weighed clause by clause, his care in deploying terms becomes apparent, and alleged inconsistencies and obscurities and irrelevancies by and large disappear. Not that Hume never makes mistakes, but rather that he tends to make them consistently, and further his mistakes are such that the ferreting of them makes the chase worth while.[29]

These lines well express the moral to be drawn from my discussion in this chapter.

[27] For the expression 'metaphysical agony' see Noxon, *Development*, p. 22 (and cf. J. G. Hamann's German translation of the Conclusion of Book I, 'Nachtgedanke eines Zweiflers', in *Sämtliche Werke*, vol. 4, ed. 1952, p. 346). D. T. Siebert, *The Moral Animus of David Hume*, Toronto 1990, p. 171, writes that 'Hume's rhetoric is understandable and meaningful' when 'viewed within the framework of his previous mental breakdown'. See also J. Mullan, *Sentiment and Sociability. The Language of Feeling in the Eighteenth Century*, Oxford 1988, p. 213; J. E. Sitter, *Literary Loneliness in Mid-Eighteenth Century England*, Ithaca, 1982 pp. 22–3 and 26ff.; Pears, *System*, p. 120, talks about 'the completeness and candour of this retractation'. P. Jones, *Hume's Sentiments*, Edinburgh 1982, p. 6, writes that Hume, in the conclusion of Book 2, gives the impression 'of half-succumbing to the intoxication of limitlessly wielding the sceptical tool'. On the other hand, see Richetti, *Writing*, pp. 196, 227ff.; and a wonderful analysis in Baier, *Progress*, pp. 15ff.

[28] L. Sterne, *The Life and Opinions of Tristram Shandy*, Book II, chapter 2.

[29] Butler, 'Hume's impressions', p. 123.

We have seen that the most puzzling aspect of Hume's discussion of the origin of the idea of space, his reference to the 'disposition' or 'manner of appearance', is also the key to the distinctness of the idea of space, an idea which includes a direct, original contribution of the mind. This is, I think, intimated by the very vagueness and ambiguity of Hume's language. In a text which, as I have shown, is otherwise clearly concerned with terminology, this vagueness cannot be easily explained as the result of carelessness – especially since Hume seems to be fully aware of it: it is, rather, to be seen as the pointer to a philosophical problem that he is trying to articulate. My suggestion is that this is the problem of how we come to know and describe the mind's 'powers and qualities', and that Hume's mentions of the 'dispositions' or 'manners of appearance' of spatial (and temporal) perceptions (like his description of belief as a 'lively', 'strong', and 'vivacious' 'manner of conception') may be seen as a form of acknowledgement, on his part, of the trace, in experience, of the mind's operations.

The special link between the self and the idea of space that I have described gives a precise and attractive sense to the distortions in the language that one finds in the treatment of the ideas of space and time: they turn into clues to the presence of the self. Conversely, the link between space and the self is not openly described either in the treatment of the origin of the idea of space, or anywhere else in Book I of the *Treatise*. All we may find about it is implicit. I shall come back to this issue in the conclusion.

INTERMEZZO

The minds of an author and his readers

> I was just going, for example, to have given you the great out-lines of my uncle *Toby*'s most whimsical character; – when my aunt *Dinah* and the coachman came a-cross, and led us a vagary some millions of miles into the very heart of the planetary system: Notwithstanding all this, you perceive that the drawing of my uncle *Toby*'s character went on gently all the time; – not the great contours of it, – that was impossible, – but some familiar strokes and faint designations of it, were here and there touch'd in, as we went along, so that you are much better acquainted with my uncle *Toby* now than you was before. By this contrivance the machinery of my work is of a species by itself; two contrary motions are introduced into it, and reconciled, which were thought to be at variance with each other. In a word, my work is digressive, and it is progressive too, – and at the same time.
>
> (*Tristram Shandy*, I, xxii).

INTRODUCTION

So far I have examined sections 1–3 of Book 1, Part 2 of the *Treatise*. These three sections contain what Hume presents as the two crucial tenets of his doctrine: that the idea of space is composed of indivisible minima, and that it is the idea of the manners of appearance of these minima. In chapter 1, I have shown how the refutation of the 'doctrine of the infinite divisibility' is crucial to the understanding of Hume's way of setting the problem of the existence of external reality in terms of the relation we may establish between mental and real objects. In chapter 2, I have suggested that we regard Hume's investigation into the origin of the idea of space in the 'manner of appearance' of perceptions as connected with the problem of the possible contribution of the mind to experience, and with our knowledge of the mind's structures and operations. These two chapters taken together outline a picture of Hume's theory of space in which the main character is the mind, with its unknown 'powers and qualities'. The essence of the mind, Hume writes, 'is equally unknown to us with that of external bodies' (*T*/XVII); so, in Part 2, I shall try to sketch a map of Hume's opinions on these unknown structures and operations by teasing information out of his sections 4 and 5, which contain his answers to possible objections. In particular, I shall focus on his treatment of the problems arising in connection with a priori knowledge and with the idea of a vacuum. But now, before examining the rest of Hume's treatment of space, I intend to pause for reflection, and justify my decision to work on the idea of space and to regard it as central to the philosophical substance of the *Treatise*.

In this intermezzo, I consider Hume's *Treatise* from a different point of view, and highlight aspects of its contents, history, and readings that are of great importance to the understanding both of Hume's treatment of the idea of space, and of ourselves as its late readers and interpreters. In particular, I shall focus on Hume's own readings of the *Treatise* in his other philosophical writings, on what he tells about it in his private correspondence, and on its early (and not-so-early) reception. My discussion will turn around two main characters: Hume and the *Treatise* itself. To each of them there corresponds a classic interpretative problem – Hume's intentions and the reactions of readers of the *Treatise* respectively – and in each interpretative problem there is the latent danger of an excess. Together with these two main characters many others will appear, most notably readers of the *Treatise*: some very scholarly philosophy professors, authors of classic contributions to Hume studies, Hume as author of public and private comments on the *Treatise*, and a trendy literary critic. Finally, comments by James Boswell, and by two anonymous reviewers of the *Treatise*, will be the focal points of my discussion of the expectations, sensibilities, conventions, and reactions of contemporary readers – the 'eye of the age', which can guide interpretation beyond the pitfalls, on the one hand, of secret authorial intentions, and, on the other, of the infinite possible meanings of written words. I shall show also that the external evidence available makes the case for the centrality of the idea of space look quite promising.

HUME'S INTERPRETATIONS OF THE *TREATISE*

In the famous advertisement to the 1777 edition of his *Essays and Treatises*, Hume writes:

Most of the principles, and reasonings, contained in this volume, were published in a work in three volumes, called *A Treatise of Human Nature*: A work which the Author had projected before he left College, and which he wrote and published not long after. But not finding it successful, he was sensible of his error in going to the press too early, and he cast the whole anew in the following pieces, where some negligences of his former reasoning and more in the expression are, he hopes, corrected. Yet several writers, who have honoured the Author's Philosophy with answers, have taken care to direct all their batteries against that juvenile work, which the Author never acknowledged, and have affected to triumph in any advantages, which, they imagined, they had obtained over it: A practice very contrary to all rules of candour and fair-dealing, and a strong instance of those polemical artifices, which a bigoted zeal thinks itself authorized to employ. Henceforth, the Author desires, that the following pieces may alone be regarded as containing his philosophical sentiments and principles.

He is here acknowledging the *Treatise*, only to ask one rather sharply not to read it (thus anticipating, as we shall see, the reaction of at least one polite reader). The only other official acknowledgement of the *Treatise* is in *My Own Life*, where Hume writes, quoting from Pope,[1] that it 'fell *dead-born from the press*'.

It is not easy to decide whether the *Treatise* or the *Enquiries* is more directly responsible for Hume's fame as a philosopher. At any rate, as a consequence of Hume's explicit disowning of the *Treatise*, in many interpretations of his philosophy there is some uncertainty about where to look for his most genuine and interesting philosophical opinions. Where is the most stimulating philosophy to be found: in the ambiguous, involuted pages of the *Treatise* or in the terse prose and thought of the *Enquiries*? In the early tour-de-force of the anatomist, or in the mature masterpiece of the painter?[2] From a different point of view, where is the real philosophy of Hume, that is, the philosophy he really meant, to be found? Related to this there is another question, which has to do with the image of Hume, the author, that one finds in the pages he has written: which Hume is the real one? Is the young, ardent, abstruse metaphysician a more or a less real Hume than the impassive, smiling 'Ambassador from the Dominions of Learning to

[1] A. Pope, *Epilogue to the Satires*, Dial. II, lines 226ff: 'All, all but truth, drops dead-born from the press, / like the last Gazette, or the last Address'.

[2] The image of the metaphysician as anatomist, with his deep and exact reasoning, while the painter stands for the moralist, with his delicate sentiment and persuasive writing, appears throughout Hume's work. See *T*/620–1, and Hume's letter to Francis Hutcheson, 17 September 1739, in Greig I, pp. 32–4, and *E*/5.

those of Conversation'?[3] This is a kind of question that may seem particularly important to the readers of the *Treatise*, where the presence of the author is so compelling and convincing that the *Treatise* itself has been called a book 'full of egoism'.[4]

These questions seem to me to lead, for various reasons, in unprofitable or frankly wrong directions. For a start, it is not clear that, if one decides that one wants to find the real philosophy of Hume, Hume himself is the best guide. It is now a matter of common wisdom that authors are not always the most reliable (and interesting) readers and interpreters of their own works.[5] More generally, I am not at all convinced that looking for the real philosophy of Hume is a good idea. In a radical mood, one may wonder how far it is justifiable to consider all the works of the same author as a whole, to use the one to illuminate the other, or even to compare them.[6] I am however a great believer in common sense (in a commonsensical sort of common sense), and I do not intend to go as far as that. So, I assume that, in the end, Hume's works

[3] The expression appears in the essay 'Of essay Writing', published in 1742 and afterwards suppressed, and then in *E*/10.

[4] Selby-Bigge, in the *Introduction* to his edition of the *Enquiries*, 2nd edition, Oxford 1902 (repr. 1972), p. ix, reports this expression as from an early review of the *Treatise*. The original has 'egotisms', for whose specific connotations in the period see below.

[5] So I disagree with T. E. Jessop, 'The misunderstood Hume', in W. B. Todd (ed.), *Hume and the Enlightenment*, Edinburgh 1974, p. 6: 'I take it for granted that he knew better than anyone else where [his] philosophy is to be found and what it is', and with R. F. Anderson, 'The location, extension, shape and size of Hume's perceptions', in D. W. Livingston and J. T. King (eds.), *Hume: A Re-evaluation*, New York 1976, pp. 153–71 (especially pp. 161 and 163). Nor do I think of the advertisement as 'the posthumous utterance of a splenetic invalid', as does T. H. Grose (in *The Philosophical Works of David Hume*, ed. by T. H. Green and T. H. Grose, 4 vols., London 1874–5, vol. 3, 'History of the editions', p. 39 n.). A more balanced position is D. T. Siebert, '"Ardour of youth": the manner of Hume's *Treatise*', in R. Ginsberg (ed.), *The Philosopher as a Writer: The Eighteenth Century*, London and Toronto 1987, pp. 177–98 (also in his *The Moral Animus of David Hume*, London and Toronto 1990, pp.170ff.).

[6] For a provocative version, see for instance R. Rorty, 'The pragmatist's progress', in U. Eco, *Interpretation and Overinterpretation*, ed. by S. Collini, Cambridge 1992, pp. 89–108: 'Eco seems to think that it was all right for Bonaparte to read 'Morella' in the light of 'Ligeia'. But why? Merely because of the fact that they were written by the same author? Is that not being unfaithful to 'Morella', and running the danger of confusing the *intentio operis* with an *intentio auctoris* inferred from Poe's habit of writing a certain sort of text?' One might indeed have fun applying to Hume and his writings his own arguments on historical testimony from the essay on miracles, in the way Richard Whateley applied them to show that Napoleon never existed (*Historical Doubts Relative to Napoleon Bonaparte*, London 1819). More sober and closely reasoned doubts on this score are to be found in Q. Skinner, 'Meaning and understanding in the history of ideas', *History and Theory* 8, 1969, pp. 3–53, especially in the section on the 'mythology of coherence'. See also J. G. A. Pocock, *Virtue, Commerce and History*, Cambridge 1985, particularly pp. 23ff. Typical products of the quest for 'unity irrespective whether the author can be shown to have intended to proceed on any such principle' (p. 24) are to be heard, according to Pocock, at the meetings of the Hume Society!

were written by Hume; that Hume wrote the *Treatise*, the *Abstract* (probably), the *Essays*, the *History of England*, and so on. And, since all these works were written by Hume, I think of them as somehow related. With this, I have made it clear that I also assume that Hume was a person, that is, something rather familiar to me, even though the reasons for this familiarity are, in turn, very unclear. It is this feeling of common humanity – the spontaneous recognition of intentionality – that organises the heap of paper into a pattern.

This last point needs some further explanation. Hume is a tricky author – his books, in particular the *Treatise*, are difficult, and offer many temptations to specious interpretation. The author, the 'I', appears as the main character of the philosophical investigation, and he certainly is a very lively and imposing presence – all the more so, since Hume is not explicitly writing a diary or an autobiography,[7] so that the 'I' makes its first appearance in the *Treatise* completely unexpectedly. I am not thinking of the 'I' of the Advertisement to the *Treatise* – 'If I have the good fortune to meet with success . . .', etc. – which is, on the whole, rather conventional, but of the one that only appears later, in *T*/1–3:

> All perceptions of the human mind resolve themselves into two distinct kinds, which I shall call IMPRESSIONS and IDEAS . . . I believe it will not be very necessary to employ many words in explaining this distinction . . . The first circumstance, that strikes my eye, is the great resemblance betwixt our impressions and ideas in every other particular, except their degree of force and vivacity . . . Upon a more accurate survey I find I have been carried away too far by the first appearance.

Here the author's 'I' pops up giving suddenly the sense of the temporality of a reflection which in the introduction had started as, on the whole, impersonal. But this is not all, for he also seems to appear at another, much more challenging level. The discussion is so multifarious, its tone so eager – Hume would certainly write 'ardent' – and, at the same time, the writing appears so perfectly balanced and tightly controlled, that it is difficult not to imagine Hume's full awareness and deliberation behind it, line by line. This is what, in different ways and to different extents, readers like Richetti and Christensen have read into Hume's works.[8]

7 I am thinking of Descartes' *Meditations* and *Discourse on Method*, as read by J. Rée, *Philosophical Tales. An Essay on Philosophy and Literature*, London and New York 1987, chapter 1, pp. 5–30.

8 J. Richetti, *Philosophical Writing*, Cambridge (Mass.) 1983; J. Christensen, *Practicing Enlightenment. Hume and the Formation of a Literary Career*, Madison and London 1987.

But with all this, the author – the organising intelligence deliberately planning the *Treatise* – is still little more than paper: his existence is a fact in a literary world. If one wants to go beyond this general feeling of a common humanity based on a literary fact, the situation soon becomes very difficult. Is it ever possible to see the real person behind the books, letters, documents, and so on; and, more generally, is it sure that there is something – some moment or intention – particularly real, more real than the rest in a human life? To put it in more neutral terms, what I intend to do is to consider Hume and to read his works following the line suggested by one of his own images of the self:

> One thought chaces another, and draws after it a third, by which it is expell'd in its turn. In this respect, I cannot compare the soul more properly to anything than to a republic or commonwealth, in which the several members are united by the reciprocal ties of government and subordination, and give rise to other persons, who propagate the same republic in the incessant changes of its parts. And as the same individual republic may not only change its members, but also its laws and constitution; in like manner the same person may vary his character and disposition, as well as his impressions and ideas, without losing his identity. (*T*/261)

This image suggests something important. Our feeling of common humanity, in its generality, does not lead interpretation very far – something more specific is as difficult as it is necessary. Now, if Hume – an author – is like a commonwealth, it is possible to adopt an eclectic attitude which, while it does not consider such documents as the *Advertisement* to the *Essays and Treatises* as the last word to be said on the *Treatise*, still finds in it, as well as in many other documents, a useful element for the understanding of the book. The resulting interpretation will not privilege any of Hume's attitudes to the *Treatise* and will not entirely rely on any of Hume's ways of telling the story. Several such attitudes and stories will, however, contribute to this account.

The first *Enquiry*, it has been written, 'is a re-cooked dish of the old "Treatise of Human Nature"'.[9] This is, after all, what is suggested in Hume's official words on the *Treatise* – in *My Own Life* (with the distinction between manner and matter in the *Treatise*), and in the way the *Treatise* is acknowledged and disowned in the *Advertisement*. What I find inconvenient in this reading is that it is based on these two texts. Both were written very late by Hume, and focus on the problem of giving the facts a unity and a direction, to organise them into a reasonably

[9] See the anonymous review of Burton's *Life and Correspondence of David Hume* (Edinburgh 1846, 2 vols) published in the *Quarterly Review* 78, 1846, Article VI, pp. 75–113 (p. 88).

consistent biographical pattern.[10] It is true that they rehearse a lesson Hume had started reciting earlier; but, indeed, the same valedictory tones also characterise those earlier documents. For example, in the letter to John Stewart, of February 1754:

> that you may see I wou'd no way scruple of owing my Mistakes in Argument, I shall acknowledge (what is infinitely more material) a very great Mistake in Conduct, viz. my publishing at all the Treatise of human nature . . . The Arguments have been laid before the World . . . human life is so short that I despair of ever seeing the Decision. (Greig 1: 187)[11]

Hume is already addressing posterity, as in the two official documents – even the stoical attitude is already there (together with the typical consideration of a 'Mistake in Conduct' as 'more material' than any possible 'Mistakes in Argument'). But, in any event, how can one maintain that the same book written in a different way remains the same book?

I suggest that we regard the *Enquiry* as a creative reading of the *Treatise*, that is, as a part of the history of its reception.[12] In turn, in so far as the *Enquiry* is a work independent of the *Treatise*, the history of its reception is perhaps to be considered as part of the history of a different style of philosophy, a style which includes the contact between philosophy and politeness, between the learned and the conversible societies (corresponding to 'argument' and 'conduct' in Hume's letter to John Stewart). But taking the *Enquiries* as an interpretation of the *Treatise* requires some caution; moreover, in so far as the *Enquiries* are considered as a reading of the *Treatise*, one must bear in mind that the history of their reception is also the history of a misunderstanding and of its growth.

There is, of course, another reading of the *Treatise* by Hume – the *Abstract*, published in March 1740. As we know from Hume's letter of the 7th March 1740 to Francis Hutcheson,[13] he had thought of having it published in *The History of the Works of the Learned*, a learned monthly serial giving abstracts and reviews of books.[14] Since, however, a review of the *Treatise* had already appeared there, in two instalments, in November and December 1739, the *Abstract* ended up being published as a booklet.

10 See E. C. Mossner, *The Life of David Hume*, Edinburgh 1954, p. 591, according to whom *My Own Life* is 'in part, autobiography and, in part, manifesto'.

11 See also the letter to Elliot of Minto of March/April 1751, Greig 1: 158, cited below.

12 This is in line with the general approach of N. Kemp Smith, *The Philosophy of David Hume*, London 1941, pp. 536ff.

13 Greig 1, pp. 37–8.

14 On *The History of the Works of the Learned*, its origin and aims, see W. Graham, *English Literary Periodicals*, New York 1930, pp. 196–226, particularly pp. 200–1 and 204–5.

Keynes and Sraffa have shown very convincingly that the *Abstract* was written by Hume.[15] I find one particular puzzling: in both the *Treatise* and the *Enquiry*, the author, in talking about mathematics, mentions algebra, arithmetic, and geometry; in the *Abstract*, only arithmetic and geometry appear. Of course, the absence of algebra in the *Abstract* may well be, say, a slip, or a misprint: these things happen all the time. But the point I intend to make is more general. The *Abstract* is a very stylish performance; Sraffa and Keynes are right: the author is a splendid writer. But not only does algebra disappear, there is also no trace of some of Hume's typical, idiosyncratic expressions. For instance, not once does the author call a crucial point under discussion 'this particular' – as Hume constantly does elsewhere.[16] Now, some idiosyncrasies – 'this particular', 'geometry, algebra and arithmetic' (or Sterne's 'by the bye') – seem to live in the pen of a writer. One is tempted to say that Hume, in order to write something like the *Abstract*, must have been well-enough aware of them to suppress them. On the other hand, by pointing out and insisting on this oddity I am not trying to cast any doubt on the attribution of the *Abstract* to Hume. I am happy to grant that the *Abstract* was written by Hume. The point is that Hume wrote the *Treatise*, Hume wrote the *Abstract*, and yet the authors of the two works – published in such a short interval – look as if they were not quite the same person. How is it that Hume has suppressed his voice in this way and has written, as it were, with another pen?

But both the *Treatise* and the *Abstract* were published anonymously – this almost invites us to pretend that we do not know that their authors, as well as that of the *Enquiries*, were the same person. Let us focus our attention on the *Abstract* and on its author. The arrangement of matters

15 *An Abstract of a Treatise of Human Nature, 1740. A Pamphlet hitherto Unknown of David Hume.* Reprinted with an Introduction by J. M. Keynes and P. Sraffa, Cambridge 1938 (repr. Hamden, Connecticut 1965). The internal evidence produced by Keynes and Sraffa has never, to the best of my knowledge, been challenged. Some doubt has been cast on the external evidence in J. O. Nelson, 'Has the authorship of the *Abstract of a Treatise of Human Nature* really been decided?', *Philosophical Quarterly* 26, 1976, pp. 82–91; but see the answers of R. W. Connon and M. Pollard, 'On the authorship of "Hume's" *Abstract*', *Philosophical Quarterly* 27, 1977, pp. 60–6, and of P. H. Nidditch, *An Apparatus of Variant Readings for Hume's Treatise of Human Nature*, Dept. of Philosophy, University of Sheffield 1976, pp. 47ff. For Nelson's more recent position and a series of (in my opinion conclusive) answers to it, see his 'The authorship of the *Abstract* revisited', *Hume Studies* 17 (1), 1991, pp. 83–6, and J. Broome, 'On the authorship of the *Abstract*: a reply to John O. Nelson', *Hume Studies* 18 (1), 1992, pp. 95–104; D. Raynor, 'The authorship of the *Abstract* revisited', *Hume Studies* 19 (1), 1993, pp. 213–16; D. F. Norton, 'More evidence that Hume wrote the *Abstract*', *ibid.*, pp. 217–22.

16 Such use of 'this particular' is, of course, by no means peculiar to Hume's writings: for instance, it is also typical of Addison.

in the *Abstract* is very clever. In order to present the new book, the author has chosen to treat in detail only one subject, the inference from cause to effect:

> Suppose I see a ball moving in a streight line towards another, I immediately conclude, that they will shock, and that the second will be in motion. This is the inference from cause to effect; and of this nature are all our reasonings in the conduct of life: on this is founded all our belief in history: and from hence is derived all philosophy, excepting only geometry and arithmetic. If we can explain the inference from the shock of two balls, we shall be able to account for this operation of the mind in all instances. (*A*/650)

The shock of the two balls is not taken from the *Treatise*; it is one of the central illustrations of causal inference in the *Enquiry*, but it appears here for the first time, presumably borrowed from Malebranche, in a Humean discussion of the idea of cause and effect. Accounting for it is prescribed as the central goal for a metaphysics aiming at the explanation of 'the operations of the understanding' with respect to

> probabilities, and those other measures of evidence on which life and action entirely depend, and which are our guides even in most of our philosophical speculations. (*A*/647)

So, the shock of the two balls is the crucial problem of metaphysics because it is the paradigm of probabilities, which are the basis of all philosophy with the sole exception of the mathematical sciences.

The author of the *Abstract* is an 'I' that appears openly in the Introduction (*A*/643–4), there to declare his intentions:

> to render a larger work more intelligible to ordinary capacities, by abridging it.

He presents this as an intrusion, for which he apologises; but one may also read it as a response to the request of the author of the *Treatise* in the *Advertisement* to the first two books, which seems positively to invite an answer of this sort:

> The approbation of the public I consider as the greatest reward of my labours; but am determin'd to regard its judgment, whatever it be, as my best instruction.

According to the author of the *Abstract*, the *Treatise* is a 'bold attempt'; it 'has been complained of as obscure and difficult to be comprehended' because of its 'length and . . . the abstractedness of the argument'; it has a remarkable 'air of singularity, and novelty'. The Author of the *Treatise* also appears as a character in the *Abstract*. He

> insinuates, that were his philosophy receiv'd, we must alter from the foundation the greatest part of the sciences;

but he 'must be contented to wait with patience for some time', for the natural audience for the *Treatise* is 'the learned world', or 'the Few', whose 'verdict' takes time. Such a verdict is also more likely to be biased than that of 'the people, who in all matters of common reason and eloquence are found so infallible a tribunal'. 'The people' are clearly 'the public', 'the Readers' to whom the *Abstract* is addressed: readers of 'ordinary capacities', 'those who are not accustomed to abstract reasoning' and hence 'are apt to lose the thread of argument, where it is drawn out to a great length', and will instead 'more easily apprehend a chain of reasoning, that is more simple and concise'. This is, I suggest, a description of the conversable society to whom Hume was to address the *Essays* and the *Enquiries*.

Though, in the Introduction, the *Abstract* is manifestly addressed to the conversable world, the text which follows makes it clear that it is also meant to create an audience for the *Treatise* in the learned world – indeed, to create the learned readers of the *Treatise* out of the group of the conversable readers of the *Abstract* itself:

> But who-ever would thoroughly understand this must consult the author himself. 'Tis sufficient, if I can make the learned world apprehend, that there is some difficulty in the case, and that who-ever solves the difficulty must say something very new and extraordinary; as new as the difficulty itself [on the idea of force or energy]. (*A*/657)

In the text the 'I' or 'we' of the author of the *Abstract* appears variously (but rather discreetly), to present the reader with a map or a summary of either the next move, or of what has already been said. The author of the *Treatise* is frequently mentioned in these same passages; they constitute, as it were, the landmarks of the story, spoken in the voice of the author of the *Abstract*. On the other hand, the voice of the author of the *Treatise* is reproduced almost in direct speech (with some terms or sentences in italics and two extensive quotations in inverted commas) in the discussion of the perceptions of the representative of mankind, 'I', 'we' or 'Adam'. And, certainly, it is this representative, in all its guises, who is the main character of the text; here it comes out clearly, while in the *Treatise* this 'I' nearly disappears because of the more imposing 'I' of the author.

A crucial feature of the reader of the *Treatise* who wrote the *Abstract* is that he reads, so to speak, pen in hand: Hume's reader is himself a writer. He is also sensitive enough to propose illustrations of the points made in the *Treatise* which are happier than those presented in the *Treatise* itself – the billiard-balls, Adam. He adds considerations of various sorts (thus anticipating some of Hume's own additions to the *Treatise*).[17] His declared aim is to make the arguments of the *Treatise* more perspicuous; while the author of the *Treatise* waits for the few to pronounce their judgements on his difficult book, the *Abstract* informs the public of the beginning of the new metaphysics. From the point of view of the declared intentions, one may feel that it is an author very similar to the author of the *Enquiries*, making here, in the *Abstract*, his first public appearance. This impression, however strange, deserves following up and testing.

The reader of the *Treatise* who has written the *Abstract* is one who can pick up immediately the most interesting thoughts expressed in the work. But – in what sense the most interesting: those that Hume most specially cherished, those that he took as the most likely to impress his audience? Those which best expressed Hume's 'new Scene of Thought'? (And, by the way, what is Hume's 'new Scene of Thought'?) Let us sketch a table of contents of the *Abstract* – in the Selby-Bigge edition, the *Abstract* is twenty pages (pp. 643–62).

Deep and abstract philosophy and the intellectual genealogy of the author of the Treatise, pp. 645–6;
Demonstration very well worked on, probability neglected (reference of the author of the *Abstract* to Leibniz), p. 647;
Copy principle, pp. 647–9;
Causation, pp. 649–57, with custom p. 652, belief pp. 652–5, necessary connection pp. 656–7;
Scepticism of the author of the *Treatise*, p. 657;
Personal identity, pp. 657–8;
Infinite divisibility, geometry and the standard of equality, pp. 658–9;
Passions, pp. 659–60 (Book 2 more easily comprehended, but very original);
Free-will, pp. 660–1 (with long quote);
The three principles of association, 'cement of the universe', as the most original part of the book, pp. 661–2.

What I find particularly important and revealing in this table of contents is that among the subjects chosen for the presentation of the *Treatise* at least two, which 'seem to be peculiar to himself, as indeed are most of his

[17] Keynes and Sraffa, *An Abstract*, pp. xxiv ff.

opinions' (*A*/657), raise – rather than solve – new, very difficult problems: personal identity and geometry, on which ''twere certainly to be wish'd, that some expedient were fallen upon to reconcile philosophy and common sense' (*A*/659). As we have seen, the author of the *Abstract* ascribes the same difficulty and novelty to another problem as well – that of the idea of force or energy – whose solution must be 'very new and extraordinary; as new as the difficulty itself' (*A*/657). Three questions left open are at the core of the presentation of the book.

It is evident that the author of the *Abstract* is a competent man of letters, who, although judging from a distance, is ready to look for the true message of the book and to recognise its good points: in particular its originality, which he seems to value most of all. This alone says a lot about him – for it is far from obvious that being very original is a good thing from the point of view of an eighteenth-century reader. He is a sympathetic reader, but he also seems, for some reason, not entirely uncritical – a bit sceptical and even a bit patronising. In short, he is a perfect reader for such a work as the *Treatise*: somebody with whom the young ardent philosopher would be delighted to engage in deep and abstruse conversation. But he is, perhaps, a fictional character. If Hume is the author of the *Abstract*, this character is a disguise, and it is Hume who, in order to attract attention on the *Treatise*, is creating his own model reader of the book.[18] He is also projecting a reader able to understand his work and to rewrite it more clearly and more attractively. Hume is, for a while, two different authors: the one writing for the few, the other translating the resulting philosophy into a style more suitable and more appealing for the public. While doing so, he chooses some subjects instead of others – among them, some which are 'abstruse and profound' and present intriguing difficulties. Later on he had the impression that this operation had lead nowhere, and took a completely different line. Instead of trying to create an audience for the ideas fully presented in the *Treatise* and waiting for the judgement of the few on the book, he ended up openly asking for the *Treatise* not to be read at all. In the meantime, he had rewritten his new Scene of Thought – whatever it was – with the pen of his writing reader, that is in a new, more elegant, and clearer form; after his training as an anatomist, he operated as a painter. But this time he also left behind the most 'abstruse and profound' pieces altogether – among them, the treatment of the idea of space.[19]

[18] Keynes and Sraffa, *ibid.*, *Introduction*, p. xi.

[19] For a more detailed comparison of the contents of the two works, see Selby-Bigge's comparative table of contents for *Treatise* and *Enquiries* in his edition of the *Enquiries*.

And yet, the treatment of the idea of space was not going to disappear altogether after the *Abstract*. In the *Enquiry*, some traces appear – radically simplified, just recognisable, the origin of the idea of extension and the problem of infinite divisibility occupy a couple of pages of the section 'Of the Academical or sceptical philosophy' (*E*/122 and 124–5).[20] It is also known that, around 1755, Hume wrote a dissertation on geometry. Lord Stanhope 'convinced me, that either there was some Defect in the Argument or in its Perspicuity: I forget which'.[21] This declaration of forgetfulness and confusion between argument and presentation is so typical of Hume – at any rate, the dissertation, which was meant to appear in the collection published in 1757 and called *Four Dissertations*, was suppressed. So, there is also Hume's ghost-reading of the *Treatise* – on the subject of space.

THE *TREATISE* AND HUME'S PRIVATE STORIES

The *Abstract* is an impressively good contemporary reading of the *Treatise*. One has no difficulty in regarding it as written by Hume himself, that is, by an author whose exceptional insight into the book he is reviewing depends not only on his being very clever, but also on his having plenty of inside information. And yet, I intend to suggest that even such a reading as the *Abstract* cannot be taken in its own right as a good guide to interpretation – and this precisely *because* it is highly likely that it had been written by Hume himself. This point may appear counterintuitive – to make it less so it may be useful now to consider once and for all the opposite, equally tricky excesses having to do with authors' intentions and readers' freedom of interpretation.

No matter whether Hume wrote it himself, or simply arranged for its publication, the insightful interpretation of the *Treatise* presented in the *Abstract* is still to be seen as an expression of Hume's initial enthusiasm; it is still an element of the first phase of a process which ends up with the rejection expressed in the *Advertisement* to the *Essays and Treatises* and in *My Own Life*. I think the story of this process is worth considering. It may be followed easily in his correspondence. Hume's letters have been preserved and published, so one can attempt to enter, so to speak, into

[20] On the other mentions of geometry in the *Enquiry* see A. Flew, *Hume's Philosophy of Belief. A Study of His First Inquiry*, London 1961; F. Zabeeh, 'Hume on pure and applied geometry', *Ratio* 6, 1964, pp. 185–191; and Flew's rejoinder, 'Did Hume distinguish pure from applied geometry?', *Ratio* 8, 1966, pp. 96–100.

[21] G. Birkbeck Hill (ed.), *Letters of David Hume to William Strahan*, Oxford 1888, p. 230.

Hume's private life (at least as private as his letters are),[22] and to follow his changes of mood and of mind quite closely. In order to compare the attitudes expressed by Hume-the man with those of Hume-the author, I cannot, I think, do better than start by piecing together the story from Hume's own words.

To Michael Ramsay, 4 July 1727:

All the progress that I made is but drawing the Outlines, in loose bits of Paper; here a hint of a passion, there a Phenomenon in the mind accounted for, in another the alteration of these accounts; sometimes a remark upon an Author I have been reading, and none of them worth to any Body and I believe scarce to my self. (Greig 1: 9)

To John Arbuthnot, March or April 1734:

Every one, who is acquainted either with the Philosophers or Critics, knows that there is nothing yet establisht in either of these two Sciences, and that they contain little more than endless Disputes, even in the most fundamental Articles. Upon Examination of these, I found a certain Boldness of Temper, growing in me, which was not enclin'd to submit to any Authority in these Subjects, but led me to seek out some new Medium, by which Truth might be establisht. After much study, and Reflection on this, at last, when I was about 18 Years of Age, there seem'd to be open'd up to me a New Scene of Thought, which transported me beyond Measure, and made me, with an Ardour natural to young men, throw up every other Pleasure or Business to apply entirely to it. (Greig 1: 13)

Same, after the description of the following crisis, the 'disease of the Learned':

Having now time and Leizure to cool my inflam'd Imaginations, I began to consider seriously, how I shou'd proceed in my Philosophical Enquiries. I found that the moral Philosophy transmitted to us by Antiquity, labor'd under the same Inconvenience that has been found in their natural Philosophy, of being entirely Hypothetical, and depending more upon Invention than Experience. Every one has consulted his Fancy in erecting Schemes of Virtue and of Happiness, without regarding human Nature, upon which every moral Conclusion must depend. This therefore I resolved to make my principal Study, and the Source from which I wou'd derive every Truth in Criticism as well as Morality. I believe 'tis a certain Fact that most of the Philosophers who

22 Both censorship and the formality of the conventions of letter writing call in question the privacy of 'private' letters in the period. See Christensen, *Practicing Enlightenment*, pp. 190ff. However, from Hume's letters it is clear that he knew how to defend his correspondence from intrusive postal inspectors (for example, see his letter of 2 December 1737 to Henry Home, in Greig 1: pp. 23–5).

have gone before us, have been overthrown by the Greatness of their Genius, and that little more is requir'd to make a man succeed in this Study than to throw off all Prejudices either for his own Opinions or for those of others. At least this is all I have to depend on for the Truth of my Reasonings, which I have multiply'd to such a degree, that within these Years, I find I have scribled many a Quire of Paper, in which there is nothing contain'd but my own Inventions . . . I found that I was not able to follow out any Train of Thought, by one continued Stretch of View, but by repeated Interruptions, and by refreshing my Eye from Time to Time upon other Objects. Yet with this Inconvenience I have collected the rude Materials for many Volumes; but in reducing these to Words, when one must bring the Idea he comprehended in gross, nearer to him, so as to contemplate its minutest Parts, and keep it steddily in his Eye, so as to copy these Parts in Order, this I found impracticable for me, nor were my Spirits equal to so severe an Employment. Here lay my greatest Calamity. I had no Hopes of delivering my Opinions with such Elegance and Neatness, as to draw to me the Attention of the World, and I wou'd rather live and Dye in Obscurity than produce them maim'd and imperfect. (Greig 1: 16–17)

Three years later, to Henry Home, 2 December 1737:

I am sorry I am not able to satisfy your curiosity, by giving you some general notion of the plan upon which I proceed. But my opinions are so new, and even some terms I am obliged to make use of, that I could not propose, by any abridgement, to give my system an air of likelyhood, or so much as make it intelligible. 'Tis a thing I have in vain attempted already, at a gentleman's request in this place, who thought it would help him to comprehend and judge of my notions, if he saw them all at once before him. I have had a greater desire of communicating to you the plan of the whole, that I believe it will not appear in public before the beginning of next winter. (Greig 1: 23–4)

To Henry Home, 13 February 1739:

'tis now a fortnight since my book was published; and besides many other considerations, I thought it would contribute very much to my tranquillity, and might spare me many mortifications, to be in the country, while the success of the work was doubtful. I am afraid 'twill remain so very long. Those who are accustomed to reflect on such abstract subjects, are commonly full of prejudices; and those who are unprejudiced are unacquainted with metaphysical reasoning. My principles are also so remote from all the vulgar sentiments on the subject, that were they to take place, they would produce almost a total alteration in philosophy; and you know, revolutions of this kind are not easily brought about. I am young enough to see what will become of the matter; but am apprehensive lest the chief reward I shall have for some time will be the pleasure of studying on such important subjects, and the approbation of a few judges. (Greig 1: 26–7)

To Pierre Desmaizeaux, 6 April 1739:

You were so good as to promise me, that, if you cou'd find Leizure from your other Occupations, you woud look over my System of Philosophy, and at the same time ask the Opinion of such of your Acquaintance as you thought proper Judges. Have you found it sufficiently intelligible? Does it appear true to you? Do the Style and the Language seem tolerable? These three Questions comprehend every thing. (Greig 1: 29)

To Henry Home, 1 June 1739, after some disappointing news about the sales of the *Treatise*:

After all, I am sensible of my folly, in entertaining any discontent, much more despair, upon this account; since I could not expect any better from such abstract reasoning; nor indeed did I promise myself much better. My fondness for what I imagined new discoveries, made me overlook all common rules of prudence; and having enjoyed the usual satisfaction of projectors, 'tis but just I should meet with their disappointments. (Greig 1: 30–1)

To Francis Hutcheson, 16 March 1740:

I wait with some impatience for a second Edition principally on Account of Alterations I intend to make in my Performance. This is an Advantage, that we Authors possess since the Invention of Printing and renders the *Nonum prematur in annum* not so necessary to us as to the Antients. Without it I should have been guilty of a very great Temerity to publish at my Years so many Noveltys in so delicate a Part of Philosophy: And at any rate I am afraid, that I must plead as my Excuse that very Circumstance of Youth, which may be urg'd against me . . . I assure you, that without running any of the heights of scepticism, I am apt, in a cool hour, to suspect, in general, that most of my Reasonings will be more useful by furnishing Hints and exciting People's Curiosity than as containing any Principles that will augment the Stock of Knowledge that must pass to future Ages. (Greig 1:38–9)

To Gilbert Elliot of Minto, March or April 1751:

I believe the philosophical Essays contain every thing of Consequence relating to the Understanding, which you would meet with in the Treatise; and I give you my Advice against reading the latter. By shortening and simplifying the Questions, I really render them more complete. *Addo dum minuo.* The philosophical Principles are the same in both: But I was carry'd away by the Heat of Youth and Invention to publish too precipitately. So vast an Undertaking, plan'd before I was one and twenty, and compos'd before twenty five, must necessarily be very defective, I have repented my Haste a hundred, and a hundred times. (Greig 1: 158)

To John Stewart, February 1754:

I shall acknowledge . . . a very great Mistake in Conduct, viz. my publishing at all the Treatise of human Nature, a Book, which pretended to innovate in all the sublimest Parts of Philosophy, and which I compos'd before I was five and twenty. Above all, the positive Air, which prevails in that Book, and which may be imputed to the Ardour of Youth, so much displeases me, that I have no Patience to review it. But what Success the same Doctrines, better illustrated and exprest, may meet with, *Ad huc sub judice lis est.* The Arguments have been laid before the World, and by some philosophical Minds have been attended to. I am willing to be instructed by the Public: tho' human life is so short that I despair of ever seeing the Decision. I wish I had always confin'd myself to the more easy Parts of Erudition. (Greig 1: 187)

Hume's last word is in the *Advertisement* published in 1777 and in the accompanying comment to William Strahan, October 16, 1775:

There is a short Advertisement, which I wish I had prefix'd to the second Volume of the Essays and Treatises in the last Edition. I send you a Copy of it. Please to enquire at the Warehouse, if any considerable Number of that Edition remain on hands; and if there do, I beg the favour of you, that you woud throw off an equal Number of this Advertisement, and give out no more Copies without prefixing it to the second volume. It is a compleat Answer to Dr Reid and to that bigotted silly Fellow, Beattie. (Greig 2: 301)

What I find particularly remarkable in this story of Hume's attitudes to the *Treatise* using the words of Hume-the-man is that, after all, it looks almost exactly like the story in the words of Hume-the-author. There is an exact correspondence between:

- the approach in the letter to Arbuthnot and the conclusion of Book 1 of the *Treatise* (ardour, depression, active and social life, reflection);
- the passage from the letter to Henry Home, February 1739, and the Introduction of the Abstract (the verdict of the few takes a long time and is likely to be biased; the Treatise is very difficult, and is a revolutionary book);[23]
- the list of questions to Desmaizeaux, and the *Advertisement* to the *Treatise* (the author is ready to learn from the public);
- the first passage from the letter to Hutcheson, and the Appendix published with Book 3 of the *Treatise* (the desire to correct, add, revise);
- the second passage from the same letter, and the choice of subjects

[23] Incidentally: the similarity between these two pages, from the private document and from the *Abstract*, is close enough to be added to the internal evidence for Hume's authorship of the *Abstract*.

presented in the *Abstract* (the value of his work in the questions it asks, even when they remain unanswered);

the 'Addo dum minuo' in the letter to Minto, and the declared aim of the *Abstract* to make the *Treatise* more comprehensible by abridging it;

the passages from the letters to Minto and to Stewart, and the *Advertisement* to the *Essays and Treatises* (the request not to look for his ideas in a juvenile work).

The only real difference between the two stories is in the way they are felt by us – for, as Hume puts it,

> If one person sits down to read a book as a romance, and another as a true history, they plainly receive the same ideas, and in the same order; nor does the incredulity of the one, and the belief of the other hinder them from putting the very same sense upon their author. His words produce the same ideas in both; tho' his testimony has not the same influence on them. The latter has a more lively conception of all the incidents. He enters deeper into the concerns of the persons: represents to himself their actions, and characters, and friendships, and enmities: He even goes so far as to form a notion of their features, and air, and person. While the former, who gives no credit to the testimony of the author, has a more faint and languid conception of all these particulars; and except on account of the style and ingenuity of the composition, can receive little entertainment from it. (*T*/97–8)

In other words, the private documents tell the same story, but more persuasively. So far, so good; and this persuasiveness, grounded on sympathy, is far from unhelpful to comprehension. It is, by the way, the key of a traditional style of interpretation, that which is represented, in the case of Hume, by the work of Burton,[24] and whose legacy was taken up and brought to splendid fruition by Mossner. There is, however, a problem with this style of interpretation: namely that indulging in such intellectual intimacy with Hume is not only a bit indiscreet – it is also anything but innocuous from the point of view of the resulting interpretation.

There is one other interpretation of Hume's letters that I find particularly clever and interesting, and also a very clear example of the dangers of a certain way of approaching the problem of Hume's – of an author's – intentions. In his book *Practicing Enlightenment. Hume and the Formation of a Literary Career*, Jerome Christensen describes Hume with the words of Pocock: the author who 'intended and effected the pro-

[24] This style is actually theorized with great clarity in the anonymous review of Burton in the *Quarterly Review*, p. 77: 'a collection like this of his letters and journals and scrap-books, setting forth the dreams and aspirations of the boy, the opinions and feelings, the loves and hatreds, the views of life, the success and disappointments of the man, all in the fresh colours and of the size and importance that nearness gives'.

duction of a body of writing systematically in accord with the postulates on which he intentionally based it'.[25] To this he adds that Hume also 'projected that body of writing in a symbolic practice, a career, that exploited, facilitated, and epitomized the operations of the commercial society which it persuasively represented'.[26] Christensen's book is biographical, but 'in a loose sense', since Hume is taken as 'a historical *figure*, the Enlightenment man of letters' (p. 5). His fundamental assumption is that 'Hume' is perfectly intentional and consistent in each of his moves as a historical figure. He also assumes that 'Hume's life can usefully be worked out in terms of his writing . . . and that both life and writing execute a strategy' – a Hume seen as truly omnipotent. The very brilliancy of Christensen's reading of Hume's correspondence and autobiographical material makes it all the more provoking. He tends, for instance, to select from Hume's correspondence and published works all the metaphors employed to talk about the *Treatise*, and to treat them as crucial pieces of evidence of some secret unease, guilt, or shame: from the 'loose bits of paper' (4 July 1727, to M. Ramsay) and the 'many a Quire of Paper' (March or April 1734, to Arbuthnot), to the more alarming idea of the body that is being castrated (2 December 1737, to Henry Home); after the publication, the beloved of whom a lover is 'besotted with a blind fondness' (13 February 1739 to Henry Home), and so on, till the final image – the stillborn child of *My Own Life*, certainly a chilling image, especially if one deliberately forgets that, after all, Hume is quoting. Even Hume's dead metaphors are called back to life. The resulting illusion of a special (and obviously privileged) insight is extremely attractive; on the other hand, the image of Hume emerging from the book is, on the whole, repulsive – for Enlightenment, of which Hume is the 'figure', has turned into bad faith, into a 'suspicious desublimation', in short into the dark side of modernity.[27] Faced with such an author, supposed to be no longer a

[25] Pocock, *Virtue, Commerce and History*, p. 25.

[26] Christensen, *Practicing Enlightenment*, p. 4.

[27] 'As far as I can tell, at their deepest reach my suspicions are bound up with an unregenerate romanticism, which rejects the hygiene of enlightenment as a program of paralysis; and I conceive of this essay as preliminary to a study of Byronic, strong romanticism which will work those prejudices into an argument. My suspicions have a more respectable pedigree out of Gramsci's theoretical profile of the traditional intellectual . . . My suspicions also owe something to the armed skepticism of Michael Foucault and to the example of Stephen Greenblatt, who has been so effective in exposing what Foucault has nicknamed the "techniques of the self" put into play by Renaissance literature' (pp. 5–6). The more general framework of Christensen's work is given by the *Dialectic of Enlightenment* of Horkheimer and Adorno (which he associates with *The Mirror of Production* of Baudrillard).

person but a 'historical *figure*', and, in addition, one towards whom he does not feel very friendly, Christensen adopts the strategy of pushing his apparently omnipotent opponent into a corner – of turning his rational omnipotence into a paranoid defence-mechanism and exposing his underlying impotence. He finds the means to do this in a psychoanalytic reading of Hume's metaphors. His story is told at that particular level of common humanity which takes the colours (apparently universal, in fact deeply characteristic of the present-day culture and sensibility alone: but, of course, Christensen knows this only too well) of a certain sort of sexual symbolism – frustrations, castration, impotence, onanism, and, above all, ghastly female-figures (Madame de Boufflers in particular) who tell you the wrong thing at the right moment or the other way round, and so on. From this point of view, the *Treatise* turns out to be 'little more than a skein of repetitions'; but this is not all, for, in fact, Hume's writings, on the whole, are 'in some measure nothing more than repetition, and his career, his historical identity, a strategy for putting himself in a position where he can repeat himself over and over again' (p. 142). What I find interesting are the grounds on which Christensen builds up this part of his interpretation:

> As we learn from the letter to Dr. Arbuthnot, the single discovery that enabled the *Treatise* was Hume's finding that 'the moral Philosophy transmitted to us by Antiquity, labour'd under the same Inconvenience that has been found in natural Philosophy, of being entirely Hypothetical, and depending more upon Invention than Experience'. That first observation came to Hume like a 'new Scene of Thought', an aboriginally impressive experience; henceforth he takes responsibility only for the mechanical repetition of that impression in 'perfectly resembling' proofs, invention after invention. (p. 142)

Here the ascribed deliberation in each move is unveiled as manic, compulsive repetition. I have expressed already my view on this: it is perfectly possible to read Hume's works after the *Treatise* as readings, or interpretations, of the *Treatise*. It is a very interesting point of view, and I also find it convincing – in the sense that it sounds right – at least as far as the *Enquiries* are concerned. However, such readings are to be considered, in my opinion, as something very different from Christensen's 'pure repetition': they are interpretations, indeed, precisely in the sense that they are (creative) misunderstandings. But this is not my point here. Let us for a moment deconstruct the deconstruction. What I find extraordinary in the passage quoted above is that Hume's 'new Scene of Thought' is identified so light-heartedly with the primordial intuition of what is, after all, nothing else but the science of human nature.

With this reading of the 'new Scene of Thought', a crucial aspect of Christensen's work becomes evident: the consciously anachronistic use of psychoanalytic techniques and the strong theoretical and ideological commitments which guide his interpretation are parasitic on traditional scholarship and its reliance on authorial intentions. His approach to the problem of interpreting the 'new Scene of Thought' is typical of the common, standard, traditional way of hunting for what authors 'really' meant. The consideration of letters and private documents, while contributing some real insight and some flavour of life, also tends to be taken too far. The 'new Scene of Thought' is a particularly good illustration of this point. This expression appears, as we have seen, in the letter of March/April 1734 to Arbuthnot.[28] It is quite beautiful, so it is frequently used to refer to Hume's philosophy in such a way as to hint at its splendid originality; I have so used it myself, and do not intend to challenge it.

This, however, is not the end of the story. For obvious reasons, many have tried to find out what exactly Hume means by 'new Scene of Thought'. Let us consider two examples. According to N. Kemp Smith, Hume perceives as the great novelty of his thought the new role attributed to feeling in the operations of the mind: the key to Hume's thought is the influence of Hutcheson, which led him 'to recognise that . . . judgments of value of whatever type, are based not on rational insight or on evidence, but solely on feeling'(p. 13); and the opening of the 'new Scene of Thought' was 'the discovery that this point of view could be carried over into the theoretical domain' (p. 13).[29] On the other hand, Passmore maintains that

> we may more than suspect that associationism was the 'new Scene of Thought', the foundation of his 'philosophical enquiries' into the moral sciences. And then it is easier than ever to understand his devoted attachment to association, the first great idea of his youth.[30]

These attempts to identify the 'new Scene of Thought' are very unsatisfactory. In particular, I doubt the presupposition that Hume is describing the moment at which he has had a specific great idea – for instance, his famous principle that ideas are copies of impressions. It is more appealing to think, with Christensen, that Hume is trying to

[28] For biographical information concerning this letter see in particular Mossner, *Life*, pp. 83 ff.; and, 'Hume's epistle to Dr. Arbuthnot, 1734: biographical significance', *Huntington Library Quarterly* 7, 1943–4, pp. 135–52.

[29] Kemp Smith, *Philosophy*, p. 13. See also pp. 17 and 20.

[30] J. Passmore, *Hume's Intentions*, 3rd edition, London 1980, p. 131.

express a sudden shift in his point of view, something he can hardly manage to put into words, an intuition he wants to develop and test; and that Hume's philosophical writings are an endless re-enactment of this experience. However, for once things may be easier than they appear – at a second reading of Hume's letter, the problem dissolves. In its context, the expression may well not refer to any particular doctrine; moreover, such an identification becomes completely uninteresting. The second state of mind, described later in the document in terms which seem to refer unambiguously to the project of a science of human nature, is usually linked to the 'new Scene of Thought' and explained in terms of 'Ardour of Youth'. But between these two parts of Hume's letter there is another crucial pair of pages which, for some reason, many readers seem to overlook in this connection, in spite of their being very famous in another. There is to be found the description of the 'disease of the Learned', that 'phenomenology of doubt'[31] which is often related to the 'metaphysical agony' of the conclusion of Book 1 of the *Treatise*.[32]

So, the *Treatise* is the outcome not of the 'Ardour of Youth' fired either by some great idea or by some primordial inspiration, but rather of both the 'new Scene of Thought' and the 'disease of the Learned', with the consequent period of reflection (during which something very close to the science of human nature appears).[33] When Hume talks of the *Treatise* as the result of the unfortunate influence of the 'Ardour of Youth', in letters written much later – I have quoted from two, written in 1751 and 1754 – he tends to identify the two stages so clearly distinguished in the early letter to Arbuthnot. He also maintains in both that the *Treatise* was 'plan'd before I was one and twenty, and compos'd before twenty five', 'compos'd before I was five and twenty' – a claim which, from the letters written in the years of the composition of the work, we know not to be accurate. In other words, in the later letters Hume is rewriting the story in the light of his present position. And we, reading the letter to Arbuthnot in the light of the later documents and hypnotised by the expression, cannot easily help being taken in by this rewritten story.

31 The expression is in J. E. Sitter, *Literary Loneliness in Mid-Eighteenth Century England*, Ithaca and London 1982, pp. 31–7 (p. 28).

32 See above, chapter 2 and fn. 27 for bibliography. In Siebert, *Moral Animus*, the connection with the letter to Arbuthnot is explicitly made and followed up.

33 In fact, to be entirely accurate we may add that while writing to Arbuthnot what Hume had in his hands was not yet the *Treatise*, but 'many a Quire of Paper'; before turning them into the *Treatise*, he was to go through two other stages – the experience of an active life in Bristol, and that of concentrated reflection in Rheims and La Flèche.

Let me add a coda to the story of the 'new Scene of Thought', that will also be a good introduction to the discussion of contemporary readings. In his *Hume's Philosophy of Human Nature*, of 1932, John Laird comments on the part of the letter about the boldness of temper and the new scene of thought, that 'in the language of the time Hume was an "esprit fort"'.[34] Now, the way Laird presents this suggestion is a bit odd. It appears in a footnote, and is followed by a reference to Heineccius' *Elementa Philosophiae* and A. Baxter's *Enquiry into the Nature of the Human Soul* – where, in the pages referred to, one finds nothing but the appearance of the term "esprit fort" applied to such people as Spinoza, Toland, Bayle, and, above all, Hobbes.[35] It is clear that Laird, who is 'anxious to make it easy for every student to check my interpretations' (p. v), thought it was the use of the term which was in need of scholarly legitimation, not the connection between it and Hume's story in the letter; in other words, what to us appears as an oddity is, I think, the sign of a different sensitivity with respect to what needs supporting evidence.

Laird hints at something about the young Hume's attitude which may help us to understand reactions like the one expressed, for instance, in Boswell's splendid assessment of the *Treatise*:

> (Saturday 28 July 1781) I borrowed today at the Advocates' Library, Hume's Treatise of Human Nature, but found it so abstruse, so contrary to sound sense and reason, and so dreary in its effects on the mind, if it had any, that I resolved to return it without reading it.[36]

Boswell's attitude may appear a bit surprising; perhaps it makes us smile a bit, condescendingly. But, in my opinion, it is not funny or silly at all. Nor is it superficial: Boswell is, I think, taking the *Treatise* deadly seriously, like a capital execution, like all that reminded him of his mortality. In his words, to read such a book is something between morbidity and bad manners (the vacillation between the two being indicated by the clause: 'if it had any'). There is something moving about the intelligent, sensitive, melancholic Boswell expressing this emphatically healthy-and-polite disapproval of philosophic excess, by deliberately and judiciously responding with this sophisticated, gossipy attitude to Hume's 'metaphysical agonies' as well as to his 'philosophy in this careless manner'. The polite way out of such doubts as those presented in

[34] J. Laird, *Hume's Philosophy of Human Nature*, London 1932, p. 4.

[35] J. G. Heineccius, *Elementa philosophiae rationalis et moralis*, Praefatio, in *Opera*, 8 vols., vol. 1, Genevae 1744, p. ix; A. Baxter, *An Enquiry into the Nature of the Human Soul*, 2nd edition, 2 vols., vol. 1, London 1737 (1st edn 1733), fn. on pp. 150ff., in particular p. 152.

[36] Quoted in M. Box, *The Suasive Art of David Hume*, Princeton 1990, p. 53.

the *Treatise* is not to adopt a different style of thought (this would mean being very original, that is, very tasteless); it is, simply, to declare one's refusal to read the book. In so doing, one also complies with Hume's explicit request in the *Advertisement* to the *Essays and Treatises*.

I find Boswell's comment revealing at different levels – revealing about Hume and the *Treatise*, and revealing about ourselves as readers of Hume, of Boswell, and of their contemporaries. We may regard Boswell's expressed opinions and stated feelings as the mirror of the polite and conversable society to which he belongs. The way he rejects free-thought and its dire effects on one's mind ('if it had any', of course) gives a sudden insight into what the *Treatise* represented to a contemporary – an ambivalent temptation, something at once intriguing and repulsive; while it also makes clear that reading the *Treatise* with the eye of the age is a difficult, perhaps impossible, task. How are we to imagine a Hume who is not *Hume*, the canonical philosopher of the *Treatise* and of the *Enquiries*, but Mr Hume, author of an excellent *History of England* and of a lamentable *Treatise of Human Nature* . . . ? Lost innocence can never be regained, and a part of this impossibility is that one cannot really *want* to regain it. The idea of following Boswell's example (as well as Hume's considered advice) that is, of failing to read the *Treatise* is not appealing. However, knowing as much as possible about Boswell's 'reading' and, in general, about the eye of the age may be extraordinarily helpful. For instance, as I have mentioned earlier, I think that if one keeps in mind that the expression 'fell *dead-born from the press*' was not an original creation of Hume to describe the unhappy story of the early reception of the *Treatise*, but a quotation from Pope, presumably rather familiar or even conventional in polite circles, then Hume's statement looses its grim and melodramatic colour; it no longer seems a conspicuous anomaly in the general smoothness and 'politeness' of Hume's late prose, but, on the contrary, appears at once very literary and very 'conversable'. The same is also true for another famous expression of *My Own Life*, when Hume calls the 'passion for Literature . . . the ruling Passion of my Life': the whole passage sounds different when one bears in mind that the Lockean expression 'ruling passion' appears, for example, in Pope's *Moral Essays* and in Roscommon's *Essay on Translated Verse*, and was, presumably, a cliché.[37]

Of course, the ability to recognise quotations, topoi, proverbs,

[37] Pope, *Moral Essays*, Ep.1, lines 174ff: 'Search then the *Ruling Passion*: there, alone / the Wild are constant, and the Cunning known; / the Fool consistent, and the False sincere; / Priests, Princes, Women, no dissemblers here.' W. Dillon, Earl of Roscommon, *Essay on Translated Verse*, 93–4: 'Examine how your *Humor* is inclin'd, / and which the *ruling Passion* of your mind.'

parodies, and so on, which were immediately evident to the contemporary reader and are now lost, is only part of what is needed. There is a point beyond which we cannot hope to anticipate the reactions of contemporaries. We need to know, for instance: what parts of the *Treatise* most attracted their attention? What in it was considered as interesting or original, and what as its good or bad points? How was the author of a book like the *Treatise* classified: as an eccentric, a genius, a clever man, an immature writer, or what? Or, moving to questions more directly related to my interest in Hume's treatment of the idea of space in the *Treatise*: how much attention was devoted to the discussion of the idea of space in the *Treatise*? Did contemporary readers regard it as more or less integrated with the rest of the book? Did they regard it as interesting, or as original? Between reconstructing Hume's deep intentions and deconstructing his texts there is, I submit, a viable medium – the reactions of contemporary readers.

A READER WITH PEN IN HAND

The public acquired information about new books through a whole class of shorter, 'secondary' writings – notices, reviews, abstracts, summaries, introductions: the reactions of the readers with a pen in hand. The image of the *Treatise* emerging from such documents is all the more important, since it is known that writings of this kind had a crucial role also in orienting and shaping the other readers' opinions and attitudes, before and during the reading itself, perhaps even instead of it (as, presumably, Boswell would have us believe in his case).[38] Also, it is with their appearance that the *Treatise* acquires its existence as a possible subject for learned and polite conversation. I have examined one such document already: the *Abstract*, which presents the *Treatise* both to the few and to the polite (ordinary) readers, and describes it as a highly original, difficult, remarkable work – a true intellectual challenge, and possibly the foundation and beginning of a new philosophy. But the

[38] For particularly fascinating materials on Hume's own use of abstracts and reviews in periodicals, see the excellent article of J. P. Pittion, 'Hume's reading of Bayle: an inquiry into the source and role of the *Memoranda*', *Journal of the History of Philosophy* 15, 1977, pp. 373–86. Cf. with the conclusions presented by M. New concerning Sterne as a reader of Locke in the *Introduction* to M. New, J. New, R. A. Davies, and W. G. Day (eds.), *The Life and Opinions of Tristram Shandy*, Gainesville 1984, 3 vols., vol. 3, pp. 1–31, particularly pp. 16ff. (see also M. Golden, 'Periodical context in the imagined world of *Tristram Shandy*', in P. J. Korshin (ed.), *The Age of Johnson. A Scholarly Annual*, New York 1987, pp. 237–60). I propose to deal with the questions of the dissemination of metaphysics and moral philosophy in eighteenth-century Britain in a further work.

Abstract was probably Hume's – and, rather than giving it a privileged role, a special relevance in the reconstruction of the reception of the *Treatise*, this should invite caution, even give rise to some suspicion about it. Now let us consider another similar document: another abstract, but this one certainly written by somebody else.

Having by these Divisions given an Order and Arrangement to his Objects (that is, I suppose, Ideas) we may now, he says, with the more Accuracy consider their Qualities and Relations. The first Circumstance that strikes his Eye*, is the great Resemblance betwixt our Impressions and Ideas in every Particular, except their Degree of Force and Vivacity. When he shuts his Eyes and thinks of his Chamber, the Ideas he forms are exact Representations (he tells us) of the Impressions he felt. I fancy most other People have made the same Observation. However, this Circumstance seems to our Author remarkable, and engages his Attention for a Moment.

*This Work abounds throughout with Egotisms. The Author would scarcely use that Form of Speech more frequently, if he had written his own Memories.

With these words, the theory of ideas of the *Treatise* is presented to the public in the only known review of the book in English – in fact an abstract, published in *The History of the Works of the Learned* in two instalments, in November and December 1739.[39] A 'somewhat abusive' document, Hume calls it; and one whose appearance caused him to publish the *Abstract* as a booklet, although he had planned to have it printed in the journal.[40] As is evident from the quotation, the article is a sharp parody based on a simple, yet irresistibly effective, rhetorical device: that of summarising closely, or even quoting extensively from, the pages of the *Treatise* – in the way which is common in the abstracts and reviews of the period – but transforming all Hume's 'egotisms', that is his uses of the first person pronouns, into third person pronouns, regardless whether they are the 'I' of the representative of humankind, the 'I' of the author sitting at his table and thinking, experimenting in real time before the reader, or the 'I' who invites the response of the reader to the book.

'Egotism' is a form of speech whose description and criticism is the subject of a whole article in Addison's *Spectator* – perhaps, the source of the reviewer's comment in the footnote.[41] Addison's article makes it perfectly clear that under the name 'egotism' a whole cluster of rhetor-

39 Pp. 353–90, and 391–404. The quotation is on p. 357.

40 See above (Letter to Francis Hutcheson, 4 March 1740; Greig 1: 38).

41 N. 562, Friday, 2 July 1714. See H. M. Bracken, *The Early Reception of Berkeley's Immaterialism 1710–1733*, 2nd edition, The Hague 1965, pp. 18ff., for further information on the eighteenth-century uses of 'egotism'.

ical and compositional moves are to be grouped, and that writing is in itself, to a certain extent, an 'egotistic' operation. 'Be present as if absent', is the Terentian motto of the piece. Commonly used by Cicero, rejected by the Port-Royal writers 'as rising from vain-glory and self-conceit', and by them given its name, egotism is the favoured form of speech in Montaigne's *Essais*, which in turn are the best illustration of egotistical writing. Egotisms also abound in books of memoirs and in 'most of our modern prefaces'. The use of egotisms is very tricky. In the case of Cicero:

How shocking soever this great man's talking of himself might have been to his contemporaries, I must confess I am never better pleased than when he is on this subject. Such openings of the heart give a man a thorough insight into his personal character, and illustrate several passages in the history of his life: besides that, there is some little pleasure in discovering the infirmity of a great man, and seeing how the opinion he has of himself agrees with what the world entertains of him.

Concerning Montaigne, the suggestion that his use of egotism may be excessive is even more open:

Had he kept his own courses, he might have passed for a much better man, though perhaps he would not have been so diverting an author. The title of an essay promises perhaps a discourse on Virgil or Julius Caesar; but when you look into it, you are sure to meet with more upon Monsieur Montaigne, than of either of them. The younger Scaliger, who seems to have been no great friend to this author, after having acquainted the world that his father sold herrings, adds these words: . . . 'For my part', says Montaigne, 'I am a great lover of your white wines' – 'What the devil signifies it to the public', says Scaliger, 'whether he is a lover of white wines or of red wines?' (pp. 28–9)

The authors of memoirs are frequently unpleasant – 'a tribe of egotists, for whom I have always had a mortal aversion', Addison writes (p. 29). In the prefaces – where, one may add, egotism was in fact standard – it may also be a bit unfortunate, for:

Every insignificant author fancies it of importance to the world, to know that he writ his book in the country, that he did it to pass away some of his idle hours, that it was published at the importunity of friends, or that his natural temper, studies and conversations, directed him to the choice of his subject. (p. 29)

The article concludes reflexively with some more deliberately and delightfully egotistic anecdotes about egotism in conversation – where, of course, it is simply bad manners.

So, in the review, the *Treatise* is censured and parodied for the use of a form of speech which, to a certain extent, cannot be avoided, but that, beyond certain limits, is distasteful and questionable. Concerning the content, the crucial point of the book is introduced by the reviewer as follows:

I do not recollect any Writer in the *English* Language who has strained a System of human Nature, morally considered, upon the Principle of this Author, which is that of Necessity, in Opposition to Liberty or Freedom. The Truth of the Principle itself has been often and very carefully discussed. Some have endeavoured to prove even the Impossibility of Liberty, while others have asserted it to be an essential Property of human Nature, the Basis of all Morality, Religion and Happiness, which can subsist upon no other Foundation, and are utterly subverted by the Denial of it.

In other words, the *Treatise* is certainly original – in the pejorative sense that it follows up recklessly a dangerous and scandalous line of thought. In the same introductory paragraph, the reviewer also presents his own canon:

To form the clearest Ideas we can have upon this abstruse Subject, we should read some Letters that passed thereupon between those acute Reasoners, Mr. *Locke* and Mr. *Limborch*, and the incomparable Dr. *Clarke*'s Answers to several Pieces of *Leibnitz* and *Collins*. (p. 353)

Locke and Clarke appear again several times in the review as the authorities who have been arrogantly overlooked in the book. The suggestion of such names to the reader invariably takes the form of a mockery of the author of the *Treatise*, for instance:

See what an extraordinary Light our Author, by two or three Arguments, has cast upon a Point, which cost Mr. *Locke*, and some other eminent Philosophers, no little Pains in settling. (p. 358, about the copy principle as a solution of the question of innate ideas)

Dr. *Clarke*, and one *John Lock*, Esq; whom he particularly names, two of the most superficial Reasoners, were, as well as many others, so weak as to fancy, that *whatever begins to exist, must have a Cause of Existence*; nay, *Hobbes* himself, as much an Atheist as we believe him, was of this Opinion. (p. 377)

Those Readers, who know all that Mr. *Lock* and his Corrector Dr. *Butler* have wrote upon this puzzling Subject, without being fully satisfied therewith, will certainly be pleas'd to find all their Difficulties vanish upon the Perusal of these few Paragraphs relating to it. (p. 390, on external existence)

Neither *Locke*, nor *Clarke*, nor the most venerable Names, shall usurp the Place of Truth in his Affections. (p. 397, about the immateriality of the soul)

At any rate, the author of the *Treatise* is not the only one of whom the reviewer disapproves. Commenting on the treatment of abstract ideas, he refers to Berkeley as follows:

It is above twenty years since I looked over that Piece of Dr. *Berkeley*'s, which contains this most precious Discovery, and, if I remember right, that Gentleman himself boasts of some mighty Advantages that would accrue from it to the Commonwealth of Learning: The Acquisition of Science was to become exceedingly easy, and several Difficulties, that were grievously to perplex Mathematicians and Metaphysicians, were to sink before it: In short, it was to do such Feast in behalf of Knowledge, as no Principle beside was able to perform. But notwithstanding all these Benefits that were to accompany it, I do not find it has met with any favourable Reception among the Literati; or that many Persons of Ability and Penetration are become Disciples: its Fortune may now perhaps be more prosperous under the Auspices of its new Patron, who, we see, undertakes to raise it above all Opposition. (p. 360)

And about the copy principle we find:

Many Wonders have been done in the Republick of Letters by a single and very simple Principle; tho' I question if any may compare with the above-mentioned, except that of *M. Leibnitz*. Every one has heard, what an immense Field of Knowledge he opened by his *sufficient Reason*, and how much wiser the World is by it, at this Day. Such Benefactors to Mankind will always be the Admiration of Posterity. (pp. 376–7)

On the whole, the article follows the *Treatise* in the order of the subjects. It consists mainly of long extracts, summaries, and quotations from the book, interspersed with openly sarcastic remarks on the author: 'The superior Capacity of our Author', 'this extraordinary Philosopher', 'so profound and accurate a Genius', and so on. The reviewer devotes 8 pages (356–64) to Part 1 (which is 7 sections, 25 pages in the Selby-Bigge edition), and 10 pages (364–74) to Part 2 (6 sections, 42 pages); the arguments of these two parts are followed thoroughly, section by section. The treatment of the idea of cause and effect, the main subject of Part 3 (16 sections, 110 pages) is more idiosyncratic: after a couple of pages on knowledge and probability (375–7), the reviewer discusses the necessity of a cause (377–81) and the two definitions of a cause in the *Treatise* (381–3); he concludes with a short reference to the rules to judge of causes and effects (383–4) and to reason of animals (384). On the whole, 9 pages. But most of his attention – for 18 pages (385–403) – is concentrated on Part 4 (7 sections, 94 pages), in particular on the problems of external existence (for 6 pages, 385–391) and of the immateriality of the soul (again for 6 pages, 393–9).

Egotisms go together with other, equally dubious devices – in particular, with a deplorable display of paradoxes.

I have remarked, that this Writer deals mightily in Egotisms; he is no less notable for Paradoxes. Some few of these I have cited, and we shall meet with many others in what follows; even in the very next, that is, the fifth section [Of the Immateriality of the Soul], which I am now entering upon, there are at least half a Dozen enough to stagger any Man who has not a strong Head-piece. (p. 393)

The consequence is the complete incomprehensibility of some parts of the book. Thus faced with 'nothing of which we can form a clear and distinct Idea is absurd and impossible', the reviewer declares: 'I have revolved this Sentence in my Mind till I have quite tired myself, but cannot, after, all, find any Meaning in it' (p. 362). Likewise incomprehensible is the rejection of the mathematical arguments for the infinite divisibility of extension (p. 367); incomprehensible the conclusion against empty space, where 'he assumes the Air of a Sphinx, only not attended with the horrible Cruelty of that Monster. He advances a Paradox at least as obscure as the other's *Enigma*' (p. 374). It is incomprehensible what the author is up to in the sections on the idea of cause and effect (p. 378), as well as his first definition of a cause, which 'lies far out of the Reach of my Capacity, and I am indeed in some Doubt whether it be intelligible to any but Men of our Author's Ability' (p. 382).

The document ends with the best-known of its passages, which turns an unfriendly review into a wonderfully ambiguous piece of writing:

Perhaps I have already and sufficiently answered the End of this Article, which is to make the Treatise it refers to more generally known than I think it has been; to bring it, as far as I am able, into the Observation of the Learned, who are the proper Judges of its Contents, who will give a Sanction to its Doctrines, where they are true and useful, and who have Authority to correct the Mistakes where they are of a different Nature; and lastly, to hint to the ingenious Writer, whoever he is, some Particulars in his Performance, that may require a very serious Reconsideration. It bears indeed incontestable Marks of a great Capacity, of a soaring Genius, but young, and not yet thoroughly practiced. The Subject is vast and noble as any that can exercise the Understanding; but it requires a very mature Judgment to handle it as becomes its Dignity and Importance; the utmost Prudence, Tenderness and Delicacy, are requisite to this desirable Issue. Time and Use may ripen these Qualities in our Author; and we shall probably have Reason to consider this, compared with his later Productions, in the same Light as we view the *Juvenile* Works of *Milton*, or the first Manner of a *Raphael*, to other celebrated Painter.[42] (p. 404)

'Somewhat abusive', Hume wrote of this article. There is also an improbable story about his towering rage – he is alleged to have visited the publisher to demand satisfaction, sword in hand. But this story was first told in 1777, that is nearly forty years after the publication of both the book and the article, and just after Hume's death, that is in a moment when the main concern of anybody writing about Hume was bound to be not so much how his philosophy had been received at its first appearance, but rather the issue of Hume's character – his atheism, and his virtuousness or depravity, stoicism or ardour, and so on.[43] So, the only clue we have to Hume's reaction is his calling the article 'somewhat abusive'. On the basis of the final paragraph, Keynes and Sraffa consider this comment as an evidence of 'the fact that authors are hard to please'.[44] On the other hand, Burton finds it characterised by 'a tone of clamorous jeering and vulgar raillery that forcibly reminds one of the writing of Warburton', the most notorious polemicist of the time.[45]

That this article is meant to be unfriendly appears clearly if one browses through the journal and has a look at the tone of the other writings it contains. In a two-year period, the only other article which is as unfriendly as that on the *Treatise*, and in the same style too, is that on Thomas Morgan's *Physico-Theology*, published in September and October 1741.[46] As Mossner has noticed, this writing is associated with the review of the *Treatise* by the editor of the journal in a note appended to its second instalment:

These Articles, on the Physico-Theology *of Dr.* Morgan, *were communicated by a Correspondent, who chuses, I find, to be concealed, with Regard to his Name, Profession, and Abode; all I can say of him is, that by the Similitude of the* MSS. *I believe him to be the Person who drew up the Account of* the Treatise of Human Nature, *which was printed in the Months of* November *and* December 1739.

The journal, which promises 'accurate' abstracts and 'impartial'

42 *Sic.* Quietly emended by Mossner, *Life*, p. 122: '. . .or other celebrated Painter'.

43 It first appeared in the *London Magazine*. See Mossner, *Life*, p. 123.

44 Keynes and Sraffa, *An Abstract*, pp. XI-XII n.; see also T. H. Huxley, *Hume: With Helps to the Study of Berkeley*, London 1897, p. 13.

45 Burton, *Life*, vol. 1, p. 109. The hypothesis of Rev. William Warburton as author of this article, advocated also by Mossner and Kemp Smith, has been convincingly challenged more recently: see R. M. Ryley, 'Did Warburton review Hume's *Treatise of Human Nature?*', *Notes and Queries* 221, 1976, 354–5. (On Warburton see A.W. Evans, *Warburton and the Warburtonians: A Study in Some Eighteenth-Century Controversies*, London 1932.)

46 Pp. 203–34 and 235–57.

reviews and abstracts of books, thus distances itself from the articles on the *Treatise* and on Morgan's book. It is interesting that, while the author of the *Treatise* is criticised for his originality and for using egotisms and paradoxes, Morgan is criticised mainly for prolixity and for being unoriginal to the point of dullness:

> Our Author has taken Care to say very few Things in this Book, which may not be found in others: Indeed he has repeated them in a way that is almost peculiar to himself: In Spirit and Style he is hardly to be parallel'd; the latter is so extremely luxuriant, that were all the Excrescencies to be pared off, the Subject of this Volume would appear in a vastly narrower Compass than it does at present; and however the Writer's or Seller's Profit might have been thereby contracted, the Reader, I am certain, would have proved no Loser. (pp. 203–4)

Accordingly, the figure of speech over-used in his work is pleonasm:

> I have cited it exactly in his own Words, that the Reader may see by this Example, how much that Figure of Speech, which Rhetoricians call *Pleonasm*, prevail in his Diction. (pp. 221–2)

Following up Burton's hypothesis, Kemp Smith and Mossner suggest that the final paragraph may have been written by another hand, that is, presumably, by the editor of the journal, in order to mitigate the general tone of the article.[47] One thing I find worth adding to their conclusions: the unambiguous reference to the youthfulness of the author of the *Treatise*, which is the crucial point of the final paragraph, seems to suggest that whoever wrote that paragraph must have had some information about the author of the book.

What attitude is this abstract suggesting to a possible reader? Should one read the *Treatise*, or not? Is such a presentation likely to have improved the sales of the *Treatise*, or the opposite? Hume found it 'rather abusive'; how did John Noon, the publisher of the *Treatise*, find it? In this abstract, the *Treatise* appears as the work of a young author, arrogant in his attitude towards Locke, Clarke, and the tradition they represent, and as abstruse as Berkeley and Leibniz; an author who uses egotisms and displays paradoxes, an atheist like Hobbes, and a sceptic. His performance is a book which is difficult, very original, intelligent, and wrong.[48] The article also implies, by the way, that the treatment of

[47] Kemp Smith, *Philosophy*, p. 522; Mossner, *Life*, p. 618.

[48] From the point of view of the sales of the *Treatise*, of course, a review of this kind may well be more of a puff than the *Abstract* itself. (For a piece of literary gossip about Hume allegedly trying to use irreligion, in vol. 1 of his *History*, as a means to attract '*popularity* and *gain*', and Hume's own version of the story, see Burton, *Life*, vol. 2, pp. 22–4, containing an extract on Hume from John Brown's *An Estimate of the Manners and Principles of the Times*, 1757, and Hume's letter of 20 May 1757 to Andrew Millar on Brown.)

the idea of space, with its egotisms, paradoxes, and implications, is wrong, scandalous, and original enough to deserve ten pages of expressed disapproval: at any rate, it is anything but marginal.

Yet again a coda is appropriate here. It is interesting that, if we have a look at the other reviews and notices published in periodicals between 1740 and 1741, it turns out that the documents in the *Works of the Learned* and the *Abstract*, between themselves, provide not only the general public, but also many of the other readers with a pen in hand, with a guide to the new book.[49] For instance, the notice published in the *Bibliothèque Britannique* sounds more or less like a plan, short but incredibly accurate, of the review in the *Works of the Learned*:

> C'est un Systême de Logique, ou plutôt de Métaphysique, aussi original qu'il se puisse, où l'Auteur prétend redresser les plus habiles Philosophes, et en particulier le fameus Mr. *Locke*, et où il avance les Paradoxes les plus inouïs, jusque à soutenir que les Opérations de l'Esprit ne sont pas libres.[50]

A similar approach also seems to guide to some extent the anonymous comment published as a letter by *The Commonsense* in July 1740, where the main objection to the *Treatise* is its adoption of 'the Doctrine of necessity'.[51] Interestingly, the defence of 'the *Liberty* of Human Action' goes on for nearly two thirds of the article – the other point discussed in the remaining third of this article is the opposition to the infinite divisibility of space. The false presupposition of the author of the *Treatise* is that 'it is a Contradiction to suppose that any *finite* Part of *Space* contains an *infinite* Number of *Parts*': on the contrary,

> tho' a certain *Magnitude* be necessary, for an *actual* Division of any Thing that can be divided, or for rendering the *Parts perceptible* to us after they are divided, yet the *Magnitude* of an Object has nothing to do with its *speculative Divisibility*, nor does the one in the least contribute to, or derogate from the other.

Now, since 'it is impossible to *divide Space* as we do *Matter*, by removing or destroying that *Contiguity* which is between any two of its *Parts*', by

49 For a list of reviews and notices of the *Treatise* see E. C. Mossner, 'The continental reception of Hume's *Treatise*, 1739–1741', *Mind* 56, 1947, pp. 31–43; *The Life of David Hume*, Edinburgh 1954, pp. 116ff.; and 'The first answer to Hume's *Treatise*: an unnoticed item of 1740', *Journal of the History of Ideas*, 1951, pp. 291–4.

50 *La Bibliothèque Britannique, ou histoire des ouvrages des sçavans de la Grande-Bretagne*, October–November 1739, p. 216. ['It is a system of logic, or rather of metaphysics, as original as possible, in which the author presumes to correct the most clever philosophers, and in particular the famous Mr Locke, and in which he puts forward the most outrageous paradoxes, to the point of claiming that the operations of the spirit are not free.']

51 *The Commonsense: or, the Englishman's Journal*, 5 July 1740.

divisibility of space one should only understand that 'every *Part* of Space, the *smallest* we can imagine, must be supposed to have *Parts*, that is to say, it must be supposed to have *Extension*'; and it is precisely the extension of every part of space which seems to be denied as a consequence of the approach of the *Treatise*. Of course, there are other reasons why the *Treatise* is an objectionable work: in particular there is the more general fact that it is a book 'abstruse and perplex'd' to the point of incomprehensibility, and thus

may impose upon weak Readers, and make them imagine, there is a great Deal of *deep Learning* in it, because they *do not understand it*.

And the article ends with the insinuation that rendering metaphysics 'obscure and perplex'd' not 'through Ignorance or Want of Capacity', but '*designedly*, for the Sake of any *private View*, or for the sake of *propagating* any *particular sect* of Religion, or any *Selfish Party* in Politicks', is 'a *most flagitious Crime*'. The author does not say what particular sect or selfish party the *Treatise* is propagating; at any rate, these words express a feeling that many have felt towards the *Treatise* and its author, and for a long time. For instance, we still find it in 1846, in the anonymous review of Burton's biography of Hume, inspiring this splendid portrait of the young sceptic:

He felt no compunctious visitings at the thought of abolishing a creed and establishing a paradox, but received his fifty pounds, and hoped to startle the world and to become a man of mark. We do not say he wrote contrary to his opinions; but to throw upon the world a book of crude unweighed philosophy, tampering in such perilous matter, is but little less criminal. Hume lived to see something of this, and to regret his juvenile performance. He was anxious that it should be forgotten, and complained of the injustice of judging him by its contents.[52]

The article devoted to the *Treatise* in the *Bibliothèque Raisonnée*, probably written by Pierre Desmaizeaux and published in 1740, is a more complicated matter. John Yolton has pointed out that this abstract is little more than a translation of the *Abstract* into French – the choice and the arrangement of the matters are identical, and most pages are, literally, translated word-by-word.[53] But the author of this document is not passively repeating the message of the *Abstract* – in fact he uses the words of the *Abstract* to express his own completely different, not very

[52] See the review of Burton's *Life and Letters* in the *Quarterly Review*, p. 83.
[53] J. Yolton, 'Hume's "Abstract" in the "Bibliothèque Raisonnée"', *Journal of the History of Ideas* 40, 1979, pp. 157–8.

positive judgement on the book. The author is clearly a highly competent professional reviewer. For a start, it is evident that he also has a copy of the *Treatise* on his desk, from which he quotes independently, takes the page numbers to be put in the footnotes, and so on; presumably he checked what he was translating against the *Treatise* itself. Here and there he interrupts the voice of the *Abstract* to add some comments of his own. For instance, just after the translation of the first few pages of the *Abstract*:

> Tout cet Ouvrage en général est rempli de pensées originales, et qui ont tout le mérite de la singularité. Peut-être trovera-t-on qu'à force de vouloir approfondir la nature intime des choses, l'Anonyme tient quelquefois à ses Lecteurs un langage peu intelligible. J'appréhende encore que ses Paradoxes ne favourisent le Pyrrhonisme. [54]

The summary just before the conclusion is another passage definitely not taken from the *Abstract*, and expresses, with some wit, a final opinion on the book as a whole:

> Il y auroit cent choses encore à remarquer sur notre Auteur, soit sur le goût de Pyrrhonisme, qui règne dans sa manière de philosopher, soit sur l'inconsistance de tant de Propositions singulières, qu'il a pris plaisir d'accumuler, soit enfin sur le conséquences pernicieuses, qu'on pourroit tirer de ses principes. Ce qu'il a de plus choquant, c'est la confiance avec laquelle il débite ses paradoxes. Jamais on ne vit un Pyrrhonien plus dogmatique. Ce n'est pas en doutant, qu'il ose substituer ses spéculations au sentiment des plus grands Philosophes sur les matières les plus abstraites, comme sur la nature de nos Idées, sur l'Etendue, sur l'Espace, sur le Vuide, sur l'Identité etc. Ce n'est pas en doutant qu'il s'avance jusqu'a soutenir, par example, qu'il est faux que tout ce qui existe doive nécessairement avoir une cause de son existence; que nous n'avons *à priori* aucune preuve de l'existence de Dieu; que les plus habiles Mathématiciens n'ont aucune certitude de la vérité de leurs Découvertes, jusqu'à ce que le Public savant y ait applaudi; et quantité d'autres Propositions aussi hardies, pour ne rien dire de plus fort. L'Auteur est sur tout cela aussi décisif qu'on puisse l'être. Les LOCKE et les CLARKE ne sont souvent à ses yeux, que des raisonneurs pitoyables et superficiels, en comparaison de lui-même; et, s'il nous est permis de parler ici son language, on voit bien que l'habitude et la coutume l'ont déjà tellement formé à croire, qu'il ne croit rien que très vivement.[55] (pp. 353–4)

54 April-May-June 1740, pp. 324–55, in particular p. 328 ['In general, this whole work is full of thoughts which are original and have all the merit of singularity. Perhaps some will find that sometimes, in order to go more and more deeply into the inner nature of things, the anonymous author addresses his readers in a language which is scarcely comprehensible. I also fear that his paradoxes may promote Pyrrhonism'].

55 ['There would be a hundred more things to notice about our author, for example, about the taste for Pyrrhonism, which prevails in his way of philosophising, or about the inconsistency of

One may observe, in passing, that the reviewer seems to be not uninterested in what the author of the *Treatise* is doing with the idea of space and related problems. At any rate, the central message is that the *Treatise* is full of paradoxes and these tend to promote Pyrrhonism; that it is an arrogant, difficult, and wrong book; and that its author deals condescendingly with such philosophers as Locke and Clarke, and is a very imposing presence in the book. Nothing to do with the *Abstract*, certainly; and yet, all this sounds, once again, rather familiar . . . And then, the final surprise:

Il est bon pourtant que le Public sache, que c'est à un jeune Philosophe qu'il est redevable des Essais dont nous venons de l'entretenir. Que ne peut-on attendre d'un génie aussi subtil et aussi profond, quand une fois l'âge aoura muri son goût, et qu'il aura eu le tems de méditer de nouveau les matières qu'il a ébauchées?[56] (p. 354)

The author of this article certainly sounds well informed about the author of the *Treatise*, even if he keeps calling him 'l'anonyme'. But what I find remarkable in this reference to the author of the book is its similarity to another English text – a paraphrase, if not a word-by-word translation into French. Even to express his personal disapproval at the end of his article, this writer is using somebody else's words: this time those of the reviewer (or the editor?) of *The Works of the Learned*.

It is now time to return to the internal connections linking the treatment of the idea of space with the rest of the work – especially, but not only, Book I – by examining the sections on 'Objections answer'd'.

so many remarkable propositions that he was pleased to pile up, or lastly about the damaging consequences which it would be possible to draw from his principles. What is most shocking is the confidence with which he delivers his paradoxes. A more dogmatic Pyrrhonian was never seen. It is not by doubting that he dares to substitute his own speculations for the opionion of the greatest philosophers on the most abstract matters, such as the nature of our ideas, extension, empty space, identity, etc. It is not by doubting that he goes so far as to claim, for example, that it is false that all that exists must necessarily have a cause of its existence; that we have a priori no demonstration of the existence of God; that the most skilled mathematicians have no certainty of the truth of their discoveries, until the learned public has approved them; and a number of further propositions which are equally daring to say the least. In all these matters the author is as assertive as one can be. In his eyes the Lockes and the Clarkes are often but pitiful and superficial thinkers in comparison with himself; and, if we are allowed here to speak his own language, we can see that habit and custom have already so much shaped him into believing, that he doesn't believe anything except in a very lively manner.']

56 ['It is good, however, that the public know that it is to a young philosopher that they are indebted for the essays on which we have just conversed with them. What can we not expect from such a subtle and deep genius, once age has ripened his taste and he has had the time to ponder again the matters he has sketched?']

PART II

Hume's 'Objections answer'd': clues to the operations of the mind

CHAPTER 3

Truth, passion, and the a priori: Section 4 and the problem of geometry

> The world-enigma in the deepest and most ultimate sense, the enigma of a world whose being is being through subjective accomplishment, and this with the self-evidence that another world cannot be at all conceivable – that, and nothing else, is *Hume's problem*. Kant, however, for whom, as can easily be seen, so many presuppositions are 'obviously' valid, *presuppositions* which in the Humean sense are included within this world-enigma, never penetrated to the enigma itself.
>
> (Husserl, *Krisis*, Part 2, section 25, pp. 95–6)

> Every part, then, of extension, and every unite of number has a separate emotion attending it, when conceiv'd by the mind.'
>
> (*T*/373)

INTRODUCTION

Part 2, section 4 of the *Treatise*, 'Objections answer'd', begins with a most welcome summary of the arguments presented in sections 1–3:

Our system concerning space and time consists of two parts, which are intimately connected together. The first depends on this chain of reasoning. The capacity of the mind is not infinite; consequently no idea of extension or duration consists of an infinite number of parts or inferior ideas, but of a finite number, and these simple and indivisible: 'tis therefore possible for space and time to exist conformable to this idea: And if it be possible, 'tis certain they actually do exist conformable to it; since their infinite divisibility is utterly impossible and contradictory. The other part of our system is a consequence of this. The parts, into which the ideas of space and time resolve themselves, become at last indivisible; and these indivisible parts, being nothing in themselves, are inconceivable when not fill'd with something real and existent. The ideas of space and time are therefore no separate or distinct ideas, but merely those of the manner or order, in which objects exist: Or, in other words, 'tis impossible to conceive either a vacuum or extension without matter, or a time, when there was no succession or change in any real existence. The intimate connexion betwixt these parts of our system is the reason why we shall examine together the objections, which have been urg'd against both of them. (*T*/40)

In the opening chapters, I have considered the two parts of Hume's 'system' and have made the case for the centrality of the self in both; now I shall proceed, in this chapter and in the next, to examine the two sets of objections and answers, through which, as I hope to show, Hume gives us some clues about the ways the mind operates. I shall start with the objections against the finite divisibility of space and against the idea of space that Hume presents in section 4.

The bulk of section 4 consists in an examination of the nature and scope of geometry, the branch of mathematics that claims to produce demonstrative arguments in favour of the doctrine of the infinite divisibility of extension. Given the kind of interest he has shown so far, it is not surprising that Hume's answer to these arguments does not involve an examination of any of them – of, say, the argument from incommensurability of the side and diagonal of a square. Rather, he considers geometry's perceptual basis, the nature of its definitions, the standard it employs in order to decide on such matters, as the straightness of a line or the equality of figures. The result is a general criticism of geometry and of its claims to certainty:

It appears . . . that the ideas which are most essential to geometry, *viz.* those of equality and inequality, of a right line and a plain surface, are far from being exact and determinate, according to our common method of conceiving them. Not only we are incapable of telling, if the case be in any degree doubtful, when such particular figures are equal; when such a line is a right one, and such a surface a plain one; but we can form no idea of that proportion, or of these figures, which is firm and invariable . . . As the ultimate standard of these figures is deriv'd from nothing but the senses and the imagination, 'tis absurd to talk of any perfection beyond what these faculties can judge of; since the true perfection of any thing consists in its conformity to its standard. (*T*/50–1)

This critical attitude to geometry is connected with the definition of knowledge as opposed to probability: later in Book 1 of the *Treatise*, in Part 3, section 1, 'Knowledge and probability', where the discussion of the status of the branches of mathematics is taken up again, Hume recapitulates his discussion of equality, and says that geometry is not really on a par, from this point of view, with arithmetic and algebra. And yet Hume still talks of geometry together with arithmetic and algebra as part of demonstrative knowledge; indeed, when he defines knowledge as opposed to probability he also takes from geometry his illustration of the certainty typical of knowledge.

In this chapter, I shall take the anomalous situation of geometry – both based on appearances, and part of certain knowledge – as an occasion for a reassessment of Hume's whole conception of knowledge and certainty. After some preliminaries on Kant and Hume, I shall examine Hume's treatment of the standard of a straight line and of the standard of equality, and his conclusion that geometry is founded on the general appearance of objects, against the background provided by contemporary mathematical works with which Hume was presumably acquainted. This will allow me to suggest a possible explanation of Hume's ambiguity concerning the status of geometry. Then, building on the link Hume establishes between demonstrative knowledge and intuitive knowledge, I shall discuss Hume's conception of certainty and the notion of a Humean a priori. Having suggested that certainty is a feature of a certain kind of acts of the mind, by way of conclusion I shall consider Hume's treatment of curiosity and the love of truth, that is, the relation between truth and the passions.

STRAIGHT LINES AND THE KANTIAN READING OF HUME

Mathematicians pretend they give an exact definition of a right line, when they say, *it is the shortest way betwixt two points*. But . . . this is more properly the discovery of one of the properties of a right line, than a just definition of it. For I ask anyone, if upon mention of a right line he thinks not immediately on such a particular appearance, and if 'tis not by accident only that he considers this property? A right line can be comprehended alone; but this definition is unintelligible without a comparison with other lines, which we conceive to be more extended.

That the straight line between two points is the shortest, is a synthetic proposition. For my concept of *straight* contains nothing of quantity, but only of quality. The concept of the shortest is wholly an addition, and cannot be derived, through any process of analysis, from the concept of the straight line.

The first of these two unexpectedly similar quotations is from the *Treatise*, the second from the *Critique of Pure Reason*.[1] The differences between them – the appeal to the reader, the 'we', the reference to appearance, intellegibility, and conception in the first, against the

[1] *T*/49–50, and *Critique of Pure Reason*, B 16, transl. by N. Kemp Smith, London 1985, p. 53. Noticed by L. W. Beck, 'A Prussian Hume and a Scottish Kant', in *Essays on Kant and Hume*, New Haven 1978, p. 118 n. 20.

precise, impersonal, objective language of the second – are emblematic of two different philosophical enterprises.

For Kantian readers of Hume, the treatment of mathematics has always been one of the main points of reference: for instance, it is the subject of the first section of the chapter on Hume in Cassirer's *Erkenntnisproblem*.[2] One of the reasons for this interest and for the direction it has frequently taken is famously to be found in Kant's texts themselves. For instance, in the *Prolegomena*, Kant writes that, according to Hume, 'mathematics contains analytical, but metaphysics synthetical, a priori propositions', and, in the *Critique of Practical Reason*, that mathematical propositions proceed 'from one determination to another only through identity, consequently in accordance with the principle of contradiction'.[3] The questions whether Kant was right in this judgement and why, and, more basically, whether the Kantian analytic and synthetic apply to Hume's treatment of knowledge, have been posed innumerable times since.

The Kantian terminology, and with it the general frame of reference for the discussion of Hume on mathematics, has been adopted by neopositivistic readers (Reichenbach and Ayer, for instance) and since then by several Hume interpreters.[4] On the other hand, Kant's reading had already been very sensibly examined and criticised at the turn of the century by the phenomenologists (in particular by Adolf Reinach).[5] Moreover, recent Kant scholars have examined the whole question again with very balanced and convincing results. In particular Beck has argued that the whole notion of certain knowledge presented by Hume has not much to do with Kant's. Thus, although the distinction in the *Enquiry* between relations of ideas and matters of fact may appear to

[2] E. Cassirer, *Die Erkenntnisproblem in der Philosophie und Wissenschaft der neueren Zeit*, vol. 2, Berlin 1907, Book 5, chapter 5, 'Hume', 1: 'Die Kritik der mathematischen Erkenntnis', pp. 248–59.

[3] I. Kant, *Prolegomena to Any Future Metaphysics*, section 4, transl. by L.W. Beck, New York 1950, pp. 17–18, and *Critique of Practical Reason*, Preface, transl. by L. W. Beck, Indianapolis 1956, pp. 13–14.

[4] See A. J. Ayer, *Language, Truth and Logic*, 2nd edition, London 1946, p. 54, and H. Reichenbach, *The Rise of Scientific Philosophy*, Berkeley and Los Angeles 1951, p. 86. See also F. Zabeeh, *Hume, Precursor of Modern Empiricism*, 2nd edition, The Hague 1973, for instance pp. 68 and 112 (for Hume, writes Zabeeh, mathematics 'is the paradigm of all sciences'); D. F. Pears, 'Hume's empiricism and modern empiricism', in D. F. Pears (ed.), *David Hume. A Symposium*, London 1963, pp. 11–30 (especially pp. 23–4); the comments of S. N. Hampshire, 'Hume's place in philosophy', *ibid.*, pp. 1–10 (especially p. 8); and A. Flew, *Hume's Philosophy of Belief*, London 1961, pp. 64ff. Cf. R. Fogelin, *Hume's Skepticism in the Treatise of Human Nature*, London 1985, p. 49: 'Hume does not, as a positivist might, identify necessity with analyticity and then turn skeptical because he is unable to find any such necessity associated with causal relation'.

[5] A. Reinach, 'Kants Auffassung des Humeschen Problems', *Zeitschrift für Philosophie und philosophische Kritik* 141, 1908, pp. 176–209 (an English translation, 'Kant's interpretation of Hume's problem', is in K. R. Merrill and R. W. Shahan (eds.), *David Hume, Many-Sided Genius*,

have something in common with the analytic/synthetic distinction, it is evident that the philosophical relations of the *Treatise* are anything but analytic in Kant's sense.[6] The Kantian distinction certainly does not apply, I would add, in the case of geometry. In fact, in my opinion, the use of such inappropriate terminology obscures rather than highlights the possible similarities between Hume's approach and Kant's.

In the first edition of the *Critique of Pure Reason*, near the beginning of the transcendental aesthetic, in the metaphysical exposition of the concept of space, one finds:

> The apodeictic certainty of all geometrical propositions, and the possibility of their a priori construction, is grounded in this a priori necessity of space. Were this representation of space a concept acquired *a posteriori*, and derived from outer experience in general, the first principles of mathematical determination would be nothing but perceptions. They would therefore all share in the contingent character of perception.

If the notion of space were linked to experience, given an experience different from that which we do in fact have, we could in principle conceive a different geometry:

> That there should be only one straight line between two points would not be necessary, but only what experience always teaches.

This link between the notion of space and experience would entail the rejection of the necessity and the universality of the notion of space, and hence of the universality and necessity of geometry. For:

Norman 1976, pp. 161–88); see also G. Preti, *Alle origini dell'etica contemporanea. Adamo Smith*, Florence 1957, pp. 56ff.; M. Dal Pra, 'L'astrazione nella filosofia di Hume', *Rivista Critica di Storia della Filosofia* 22 (4), 1967, pp. 435–56; M. Malherbe, *Hume*, Paris 1976; G. Davie, 'Edmund Husserl and "the as yet, in its most important respect, unrecognized greatness of Hume"', in G. P. Morice (ed.), *David Hume, Bicentenary Papers*, Edinburgh 1977, pp. 69–76, and D. Livingston, *Hume's Philosophy of Common Life*, Chicago 1984, chapter 2, section 3, pp. 44–59. (On Husserl's reading and use of Hume see also J. Passmore, *A Hundred Years of Philosophy*, Harmondsworth 1968 (1st edn London 1957), pp. 188ff.; and the more extensive R. A. Mall, *Experience and Reason. The Phenomenology of Husserl and its Relation to Hume's Philosophy*, The Hague 1973; R. T. Murphy, *Hume and Husserl. Towards Radical Subjectivism*, The Hague 1980; and R. Donnici, *Husserl e Hume: per una fenomenologia della natura umana*, Milan 1989.)

[6] See Beck, *Essays*, especially chapter 5: 'Analytic and synthetic judgments before Kant', pp. 80–100. For very sensitive, stimulating observations see D. Livingston, *Hume's Philosophy*, pp. 49ff. See also R. F. Atkinson, 'Hume on mathematics', *Philosophical Quarterly* 10, 1960, pp. 127–137, and the excellent essays by D. Gotterbarn, 'Kant, Hume and analyticity', *Kant-Studien* 65, 1974, pp. 274–83, and W. A. Suchting, 'Hume and necessary truth', *Dialogue* 3, 1966–7, pp. 47–60. But the Kantian terminology keeps popping up in readings of Hume – for instance: 'it is convenient to make use of some Kantian distinctions which are implicit in Hume. For Hume, all meaningful propositions are either *a priori* and analytic or *a posteriori* and synthetic' (in p. 125 of the interesting paper of B. Rollin, 'Hume's blue patch', *Journal of the History of Ideas* 33, 1971, pp. 119–28).

What is derived from experience has only comparative universality, namely, that which is obtained through induction. We should therefore only be able to say that, so far as hitherto observed, no space has been found which has more than three dimensions.[7]

In the *Treatise* there is a direct answer to the *Critique*:

I wou'd fain ask a mathematician . . . How can he prove to me, for instance, that two right lines cannot have one common segment? Or that 'tis impossible to draw more than one right line betwixt any two points? Shou'd he tell me, that these opinions are obviously absurd, and repugnant to our clear ideas; I wou'd answer, that I do not deny, where two right lines incline upon each other with a sensible angle, but 'tis absurd to imagine them to have a common segment. But supposing these two lines to approach at the rate of an inch in twenty leagues, I perceive no absurdity in asserting, that upon their contact they become one. (*T*/51)

When it [Geometry] determines, that right lines cannot concur; that we cannot draw more than one right line between two given points; its mistakes can never be of any consequence. (*T*/72)

Hume's approach sounds genuinely subversive: we do not really know that two straight lines cannot concur; we do not really know that only one straight line can be drawn between two given points – rather, we can take it that this is grosso modo the case, and can rely on the fact that we cannot go very badly wrong. True, the presence of a synthetic moment is evident; and yet, it is not at all the main point emerging from these passages: for Hume's view of the matter turns out to be a radical alternative to Kant's. In order to appreciate this fully, however, it may be useful now to offer a more systematic and detailed presentation of the discussion of geometry in the *Treatise*.

(VARYING) STANDARDS

The main argument of Part 2, section 4 consists in the presentation of the objection against Hume's perceptual indivisibles that is constituted by the existence of mathematical demonstrations implying infinite divisibility:

There have been many objections drawn from the *mathematics* against the indivisibility of the parts of extension; tho' at first sight that science seems rather favourable to the present doctrine; and if it be contrary in its *demonstrations*, 'tis

[7] *Critique of Pure Reason*, A24; N. Kemp Smith (ed.), *Dialogues*, pp. 68–69.

perfectly conformable in its *definitions*. My present business then must be to defend the definitions, and refute the demonstrations. (*T*/42)

The definitions referred to here are the common ones of surface, line, and point:

a surface is defin'd to be length and breadth without depth; A line to be length without breadth or depth: A point to be what has neither length, breadth nor depth . . . all this is perfectly unintelligible upon any other supposition than that of the composition of extension by indivisible points or atoms. How else cou'd any thing exist without length, without breadth, or without depth? (*T*/42)

Of course, Hume reports, some other answers have been tried. For instance, some have claimed that geometrical entities do not and cannot exist, as Bayle reports in his *Dictionary*;[8] or that they may be considered as abstract entities only conceived through a distinction of reason, as Hume could find, for example, in the *Art de penser*.[9] But, in the *Treatise*, both these attempts to avoid the problem are mentioned only to be quickly dismissed as absurdities.

The point of Hume's argument, however, is not to suggest that geometry is contradictory. The claim that infinite divisibility is proved is regarded as the result of the mad pride of those mathematicians who

pretend, that those ideas, which are their objects, are of so refin'd and spiritual a nature, that they fall not under the conception of the fancy, but must be comprehended by a pure and intellectual view, of which the superior faculties of the soul are alone capable. (*T*/72)

This pretence is absurd: geometry is, Hume maintains, grounded on perception. The contrast between geometrical definitions and demonstrations arises from asking it questions that it is not equipped to answer:

none of these demonstrations can have sufficient weight to establish such a principle, as this of infinite divisibility; and that because with regard to such minute objects, they are not properly demonstrations, being built on ideas, which are not exact, and maxims, which are not precisely true. (*T*/45)

The point is, as I have already shown, that the minima of perception

8 For instance, in the footnotes to the entries 'Kepler', 'Hobbes', 'Zeno Eleatic', and 'Zeno Epicurean'.

9 A. Arnauld and P. Nicole, *La Logique ou l'Art de Penser*, ed. by P. Clair and F. Girbal, Paris 1965, Part 1, chapter 5. And see I. Barrow, *The Usefulness of Mathematical Learning Explained and Demonstrated; Being Mathematical Lectures Read in the Public Schools at the University of Cambridge* by Isaac Barrow, D.D. Professor of Mathematics, and Master of the Trinity College, etc. Translated by the Revd Mr John Kirkby, of Egremont in Cumberland, London 1734 (facsimile edition, London 1970), p. 137.

are far too elusive to provide a good foundation for demonstrative knowledge. (As we shall see in a moment, this is stated even more explicitly in connection with the attempt to found an exact definition of geometrical entities on the minima of perceptions.)

So the supposed geometrical demonstrations of the doctrine of infinite divisibility do not, in principle, differ from the 'scholastick quibbles' (*T*/32) or the 'heap of unintelligible cavils and distinctions' (*T*/44) displayed by swollen-headed schoolmen, and the fundamental ideas of geometry are not 'abstract and refin'd perceptions'. Indeed, they appear to be concrete perceptual objects. How does one come to conceive a straight line as opposed to a curved one? Echoing Malebranche, Hume writes:

> Nothing is more apparent to the senses, than the distinction betwixt a curve and a right line; nor are there any ideas which we more easily form than the ideas of these objects. But however easily we may form these ideas, 'tis impossible to produce any definition of them, which will fix the precise boundaries betwixt them;[10] (*T*/49)

and also:

> When we draw lines upon paper or any continu'd surface, there is a certain order, by which the lines run along from one point to another, that they may produce the entire impression of a curve or a right line; but this order is perfectly unknown, and nothing is observ'd but the united appearance. (*T*/49)

A straight line is perceived as such immediately, from its 'united appearance'. The technical definition of a straight line – the *Treatise* mentions that according to which it is 'the shortest way betwixt two points'[11] – does not seem to have anything to do with the formation of the idea of a straight line:

> In common life 'tis establish'd as a maxim, that the straightest way is always the shortest; which wou'd be as absurd as to say, the shortest way is always the shortest, if our idea of a right line was not different from that of the shortest way betwixt two points. (*T*/50)

The straightness of a straight line is its manner of appearance: our judgement is based directly on 'an entire impression', on the spontaneous recognition of shape which occurs naturally and, as it were,

10 See *Recherche de la Vérité*, Book 1, chapter 7, division 3.

11 It is the Archimedean definition, very commonly used in place of, or together with, the famously rather unhelpful Euclidean one according to which 'a straight line lies equally (*ex aequo*) between its points' (see for example J. Ralphson, *A Mathematical Dictionary*, London 1702, s.v. 'Right line', defined as 'that which lies equally between its Points, or according to some, the shortest that can be drawn between two Points').

before thought: 'The original standard of a right line is in reality nothing but a certain general appearance' (*T*/52). Of course, Hume adds, whenever necessary the simple, primitive impression may be corrected by more accurate consideration or by comparison with some rule

> of whose rectitude from repeated trials we have a greater assurance. 'Tis from these corrections, and by carrying out the same action of the mind, even when its reason fails us, that we form the loose idea of a perfect standard to these figures, without being able to explain or comprehend it. (*T*/49)

This process of comparison initiates a kind of inertial movement of the mind towards the notion of a perfect standard. In this quest for certainty, in order to check the original impression about the straightness of this particular straight line, the mind compares it with a rule whose straightness is, in principle, no more certain, but simply commands more confidence – 'assurance' – because of the 'repeated trials'. I shall show later how important this aspect of the story is.

A 'loose idea of a perfect standard', Hume writes: and the expression itself suggests that we have to do with something whose conception is the confused product of a mind pushing itself to the limit. On the other hand, this perfect standard, the idea of which is so loosely conceived, is terribly important – if it exists, and can be found, it is the guarantor of geometrical certainty. And what if it does not exist, or cannot be found?

One comes closer to an answer from Hume's discussion of the standard of equality. What do mathematicians mean, 'when they say one line or surface is EQUAL to, or GREATER, or LESS than another' (*T*/45)? Hume aims at showing that this question is far from straightforward. To start with, the theory of minima seems to provide an obvious answer:

> Lines or surfaces are equal, when the numbers of points in each are equal; and . . . as the proportion of the numbers varies, the proportion of the lines and surfaces is also vary'd. (*T*/45)

But this is only apparently correct – it is yet again the very minuteness of perceptual minima that makes them unsuited to the purpose:

> the points, which enter into the composition of any line or surface, whether perceiv'd by the sight or touch, are so minute and confounded with each other, that 'tis utterly impossible for the mind to compute their number . . . such a computation will never afford us a standard, by which we may judge of proportions. (*T*/45)[12]

[12] See above, chapter 1, 'Levels of perception and the absolutely minute', for a fuller discussion of this passage.

It goes without saying that the theory of infinite divisibility does not allow any such standard of equality at all.

A clue to the sense of this point-by-point standard of equality appears when one is told that the same argument which pre-empts its effectiveness also applies to congruence. For Hume writes:

> if [equality] consists . . . in this imaginary application and mutual contact of parts, we must at least have a distinct notion of these parts, and must conceive their contact. Now 'tis plain, that in this conception we wou'd run up these parts to the greatest minuteness, which can possibly be conceiv'd; since the contact of large parts wou'd never render the figures equal. But the minutest parts we can conceive are mathematical points. (*T*/46)

In this connection, Hume mentions Isaac Barrow's *Mathematical Lectures*, which contain an extensive and clear account of congruence. As is known from Kemp Smith, in Barrow's Lecture XI, 'Of the congruity and equality of magnitudes', Hume read that 'congruity' is 'the Chief Pillar and principal Bulwark of all the Mathematics'.[13] The lecture goes on to identify five kinds of congruity, and these greatly help to make sense of Hume's whole approach to the standard of equality.

The first two kinds of congruity are, Barrow says, perfect congruity – discoverable between entitites of the same kind, a straight line with a straight line or a plane surface with a plane surface – and possible congruity – which connects, for instance, straight lines with the perimeters of rectilinear polygons. In the third place one finds congruity when 'all the Indivisibles of each Magnitude may succeed in the same Place, and neither of the two vary the Position of their Parts' (p. 193). The example clarifies what he means by 'indivisibles' at this point. We have this kind of congruity, he says, when we rotate a circle over a straight line, in which 'all the Points of the circular Periphery are continually applied in a successive Order to all the Points of the Right Line', or when we rotate a cylinder over a plane surface, where

> all the parallel Right Lines, which lie disposed one after another in the Superfice of the Cylinder, are applied, by a continued uninterrupted Series, to all the parallel Right Lines in the plane Superfice. (p. 194)

The 'indivisibles' are 'all the Points' of the circumference and of the right line, and 'all the parallel Right Lines' on the surface of the cylinder: Barrow is talking in the terms of Cavalieri's method of indivisibles,

[13] I. Barrow, *Mathematical Lectures*, p. 185. See N. Kemp Smith, *The Philosophy of David Hume*, London 1941, pp. 343–6.

a simple and very popular way of calculating areas and volumes, whose basis, he says, is indeed congruity.[14] According to this method, one compares areas on the basis of 'all their lines', that is of the innumerable straight lines which 'compose' them. The lines are the 'indivisibles' of the compared areas – indivisibles which are very paradoxical entities, since they are, in Cavalieri's own terms, the 'indivisibles of continuous magnitudes'.[15] Barrow's fourth kind of congruity occurs when

> all the Indivisibles of both Magnitudes do occupy the same Place in the same Time, the Situation of the Parts of one being in some measure unvaried, but yet their Order preserved, and the former Contiguity of the particular Parts retained. (p. 194)

What follows as an example of this kind of congruity is indeed a very clear illustration of Cavalieri's method:

> to this mode also belongs, or is very near of kin to it, that Congruity of which equal Figures included within the same parallel Lines or Planes are capable; of which all the Lines or Planes intercepted between the same Parallels are equal. (p. 196)

Here Barrow refers directly to Cavalieri's works for more detail. Finally there is another reference to the so-called second method of indivisibles, where the magnitudes to be compared are, as it were, chopped up:

> The fifth Way is, when the Congruity may be so performed, that the Position and Order of certain Parts is changed. This Way of Congruity, of all the most imperfect and difficult to be understood, agrees with homogeneous Figures altogether unlike to one another. Ex. gr. A Triangle is no otherwise congruent with a Circle, a Cone with a Sphere, than by transposing the Parts of either, by applying a Part of one to a Part of the other, and a Part of the Residue in one to a part of the Remenant in the other: and so perpetually, until the Business be finished, and all the Parts of the one be at length applied to all the Parts of the other. (p. 197)

The similarity between Barrow's introduction of the basic notions of Cavalieri's indivisibles and Hume's criticism of the standard of equality hardly needs emphasising. Hume's discussion, therefore, is not an idiosyncratic performance, but is in line with the treatment to be found in competent university textbooks – such as Barrow's *Mathematical*

[14] *Ibid.*, pp. 186–7.

[15] The very title of Cavalieri's classic work is *Geometria indivisibilibus continuorum nova quadam ratione promota* [that is, 'Geometry advanced in a new way by the indivisibles of the continua'].

Lectures, as well as the other mathematical source he quotes, Malezieu's manual, where he could find a more detailed, but similar and, again, very standard introduction of Cavalieri's indivisibles.[16]

As I have said, according to Hume the problem of establishing the equality of figures is soluble in principle, but intractable in practice. The idea of a perfect geometrical standard of equality is the useless and incomprehensible outcome of the same natural fiction of the mind that we have seen operating in the case of the straight line:

> We are sensible, that the addition or removal of one of these minute parts, is not discernible either in the appearance or measuring; and as we imagine, that two figures, which were equal before, cannot be equal after this removal or addition, we therefore suppose some imaginary standard of equality, by which the appearances and measuring are exactly corrected, and the figures reduc'd to that proportion. This standard is plainly imaginary. (*T*/48)

Again, the only standard we actually employ in judging the proportions of figures is that of 'a particular appearance corrected by juxtaposition or a common measure' (*T*/48); one uses a stricter or looser method of comparison according to the degree of precision that the question at hand requires. In sum, as Hume puts it in the Appendix:

> To whatever side mathematicians turn, this dilemma still meets them. If they judge of equality, or any other proportion, by the accurate and exact standard, *viz.* the enumeration of the minute indivisible parts, they both employ a standard, which is useless in practice, and actually establish the indivisibility of extension, which they endeavour to explode. Or if they employ, as is usual, the inaccurate standard, deriv'd from a comparison of objects, upon their general appearance, corrected by measuring and juxta position; their first principles, tho' certain and infallible, are too coarse to afford any such subtile inferences as they commonly draw from them. (*T*/638)

But if, in geometry, equality is only such as long as it is detectable and can be judged by the mind, the case is different with numbers:

> we are possest of a precise standard, by which we can judge of the equality and proportions of numbers . . . when two numbers are so combin'd, as that the one has always an unite answering to every unite of the other, we pronounce them equal. (*T*/71)

The geometric standard of equality is as loose and uncertain as the standard of a straight line. But, concerning the straight line, Hume was

[16] Barrow's *Lectiones* had been first published in Latin in 1683, and Malezieu's *Elemens* in 1705; both were still popular well into the eighteenth century (indeed the English translation of Barrow's *Lectiones* was first published in 1734, and Malezieu's *Elemens* were reprinted in 1729). On Barrow and his *Mathematical Lectures* see E. A. Burtt, *The Metaphysical Foundations of Modern*

ready in the end to allow that the lack of exactness 'can never be of any consequence'; on the contrary, in the case of the standard of equality there is at least one very serious consequence:

> there remain . . . algebra and arithmetic as the only sciences, in which we can carry on a chain of reasoning to any degree of intricacy, and yet preserve a perfect exactness and certainty. (*T*/71)

For geometry,

> or the *art*, by which we fix the proportions of figures; tho' it much excels both in universality and exactness, the loose judgments of the senses and imagination; yet never attains a perfect precision and exactness. (*T*/70–1)

In other words, it appears that the distinction between probability and knowledge cuts across mathematics, demoting geometry below arithmetic and algebra.

A MODEL OF KNOWLEDGE?

It is well known that, in the *Enquiry*, Hume did not repeat this criticism of geometry. Indeed, the mentions of geometry there are few and restrained. The most quoted one occurs as an illustration in the introduction of the distinction between relations of ideas and matters of fact; certain propositions

> are discoverable by the mere operation of thought, without dependence on what is anywhere existent in the universe . . . *That the square of hypothenuse is equal to the square of the two sides*, is a proposition which expresses a relation between these figures . . . Though there never were a circle or a triangle in nature, the truths demonstrated by Euclid would forever retain their certainty and evidence. (*E*/25)

Science, London 1980, pp. 150–61, and, for more recent contributions, the articles in M. Feingold (ed.), *Before Newton. The Life and Times of Isaac Barrow*, Cambridge 1990. I cite Malezieu from *Elemens de Geometrie de Monseigneur le Duc de Bourgogne*, Troisième edition, Paris 1729. Material of interest for my discussion appears especially in Books 8 and 10. More detail on Malezieu's book as a source for Hume is to be found in Kemp Smith, *Philosophy*, pp. 339–42, and in my 'Some aspects of Hume's conception of space', *Studies in History and Philosophy of Science* 21 (3), 1990, pp. 371–411. For detailed discussions of Cavalieri's methods see E. Giusti, *Bonaventura Cavalieri and the Theory of Indivisibles*, Rome 1980; and K. Andersen, 'Cavalieri's method of indivisibles', *Archive for the History of the Exact Sciences* 31, 1984–5, pp. 291–367; on their reception and fortune see especially E. Giusti, 'Dopo Cavalieri. La discussione sugli indivisibili', in L. Grugnetti and O. Montaldo (eds.), *La Storia delle Matematiche in Italia. Atti del Convegno 29 Settembre-1 Ottobre*, Cagliari 1984, pp. 85–114. See also M. E. Di Stefano and M. Frasca-Spada, 'Metodi classici e trattazione del continuo: alcuni aspetti della storia degli indivisibili', in C. Maccagni and P. Freguglia (eds.), *La storia delle scienze*, Rome 1989, pp. 158–92.

This passage is usually taken as an example of the difference between the *Treatise* and the *Enquiry*, and of the better organisation of the latter: there is here no mention of the standard of equality, and geometry is fully restored to its proper place together with the other mathematical disciplines, as one of the forms of demonstrative knowledge resulting from the relations of ideas.[17] This is, in my opinion, an interesting myth.

Let us consider again the question of equality in the *Treatise*. I have already shown that the standard of equality is based on the general appearance of objects, checked and made more accurate in a variety of ways whenever necessary. But, in the *Treatise*, we also find:

> 'Tis from the idea of a triangle, that we discover the relation of equality, which its three angles bear to two right ones; and this relation is invariable, as long as our idea remains the same. (*T*/69)

This passage is the only example of knowledge presented in the *Treatise* to illustrate the distinction between knowledge and probability; and it sounds very similar indeed to the famous page of the *Enquiry* quoted above. In both passages, the possibility of talking about equality as if it were perfect is taken for granted: equality is demonstrated with an argument, rather than measured on the basis of any concrete standard. So, in the *Treatise*, geometry is imprecise and unreliable, and at the same time the paradigm of knowledge. Now, this is a genuine inconsistency. It cannot be explained away – but one can go a considerable way towards making sense of it.

Barrow's (and Malezieu's) texts provide a helpful background for this. It is evident, I submit, that by 'geometry' Hume refers to two different 'geometries': one is 'the truths demonstrated by Euclid' (for instance, the theorem of Pythagoras, or that the angles of a triangle equal two right angles), and the other is the measure of areas of arbitrary figures (as in Cavalieri's method). From this point of view, Hume's inconsistency may, I believe, be seen as an expression of unease, or even as a criticism, however implicit and convoluted, of the attempt to export the standard of certainty of the 'demonstrations' of one 'geometry' into the territory of another, very different 'geometry'.

Yet this cannot be the whole story. In the *Enquiry*, Hume says that some aspects of geometry cast a disquieting shadow on philosophy as a whole:

[17] See for instance J. Noxon, *Hume's Philosophical Development*, Oxford 1973, pp. 114ff.; J. Wright, *The Sceptical Realism of David Hume*, Manchester 1983, p. 98; A. Flew, *Hume's Philosophy*, pp. 61ff.; R. Fogelin, 'Hume and Berkeley on the proofs of infinite divisibility', in his *Philosophical Interpretations*, Oxford 1992, pp. 45–69 (especially pp. 54–5).

The demonstration of these principles seems as unexceptionable as that which proves the three angles of a triangle to be equal to two right ones, though the latter opinion be natural and easy, and the former big with contradiction and absurdity. Reason here seems to be thrown into a kind of amazement and suspence, which, without the suggestions of any sceptic, gives her a diffidence of herself, and of the ground on which she treads. She sees a full light, which illuminates certain places; but that light borders upon the most profound darkness. And between these she is so dazzled and confounded, that she can scarcely pronounce with certainty and assurance concerning any one object. (*E*/157)

It is interesting that the equality between the angles of a triangle and two right angles, the illustration of relations of ideas in the *Treatise*, appears again as an ideal example of certainty in this passage of the *Enquiry*.[18] Here, though, the other case under discussion is not the measure of areas, but the angle of contact between the circumference and the tangent – we cannot deny that such an angle 'is infinitely less than any rectilineal angle' and yet 'as you may increase the diameter of the circle *in infinitum*, this angle of contact becomes still less, even *in infinitum*'. The demonstrative argument concerning this paradoxical entity is quite as unexceptionable as that concerning the angles of a triangle. The problem of the angle of contact is similarly and even more explicitly posed in the *Treatise*:

I desire therefore our mathematician to form, as accurately as possible, the ideas of a circle and a right line; and I then ask, if upon the conception of their contact he can conceive them as touching in a mathematical point, or if he must necessarily imagine them to concur for some space. Whichever side he chuses, he runs himself into equal difficulties . . . If he says, that in his conception of the contact of those lines he must make them concur, he thereby acknowledges the fallacy of geometrical demonstrations, when carry'd beyond a certain degree of minuteness; since 'tis certain he has such demonstrations against the concurrence of a circle and a right line; that is, in other words, he can prove an idea, *viz.* that of concurrence, to be *incompatible* with two other ideas, *viz.* those of a circle and right line; tho' at the same time he acknowledges these ideas to be *inseparable.* (*T*/53)

What Hume writes about the angle of contact is not very far from what he could find on such matters in, say, Malezieu's manual:

L'esprit humain est convaincu de certaines vérités: mais il est obligé d'avouer sa faiblesse, quand il veut comprendre, pour ainsi dire, le comment: c'est-à-

[18] And not only in Hume: for instance, in Chambers' *Cyclopaedia* (1738), in the entry 'Certitude', the equality of the three angles of a triangle to two right ones is given as the illustration of metaphysical certitude.

dire, comment il est possible que ces vérités subsistent ensemble? Mais comme l'esprit humain est borné, et que le Createur de nos ames, ne leur a pas donné des lumieres infinies, c'est à nous de nous souvenir de notre condition. Rien ne seroit plus déraisonnable, que de vouloir nier des vérités dont nous sommes convaincus d'aillieurs, parce que nous n'en comprenons pas la liaison. Nous les comprenons ces vérités, parce que nous avons une certaine mesure de raison; nous n'en comprenons pas la liaison, parce que nous ne sommes pas Dieu, et que notre raison n'est pas infinie. On a donc grand tort de vouloir attaquer la Geometrie des Infiniment Petits, et celle des Indivisibles, parce qu'il y a de certaines choses qu'on ne comprende pas dans la nature de l'infinie, qui en effet doit être incomprehensible; mais autre chose est de la comprendre, autre chose se convaincre qu'il existe. J'avoue de bonne foi, que je suis pleinement convaincu de la vérité de la deuzième proposition; mais j'avoue que je ne la comprends pas. Que si je me vois obligé de reconnoître des vérités incompatible en Geometrie, où l'esprit humain se pique de voir plus claire qu'allieurs; à plus forte raison dois-je avoir de la soumission pour des vérités d'un ordre superieur à ma raison, et me souvenir toujours que celui qui l'a créée n'étoit pas obligé de la rendre capable de tout.[19]

The same kind of argument appears again in Malezieu's Book 9, in the 'Réflection sur les incommensurables' cited in the Treatise,[20] which follows the demonstration of the incommensurability of the diagonal and the side of the square (Prop. 4):

Voila notre raison réduite à d'étrange extremités. La Geometrie nous démontre la divisibilité de la matiere à l'infini, et nous trouvons en même temps qu'elle est composée d'indivisibles. Humilions-nous encore une fois, et reconnoissons qu'il n'appartient pas à une créature, quelque excellente qu'elle

[19] Malezieu, *Elemens*, Book 11, pp. 30–1 ['The human mind is convinced of certain truths; but it is forced to admit its weakness when it wishes to understand, as it were, how: in other words, how is it possible that these truths can co-exist? But as the human mind is limited, and as the Creator of our souls has not given them infinite light, we must be mindful of our condition. Nothing would be more unreasonable than to want to deny truths about which we are otherwise convinced, because we cannot understand the connection between them. We understand these truths because we possess a certain degree of reason; we do not understand the connection because we are not God and because our reason is not infinite. It is therefore quite wrong to wish to attack the Geometry of the Infinitely Small and that of Indivisibles, because there are certain things in the nature of infinity which are not understood and which must in effect be incomprehensible. But it is one thing to understand infinity and another to convince oneself that it exists. I admit in good faith that I am fully convinced of the truth of the twelfth proposition; but I admit that I do not understand it. For if I consider myself obliged to recognise incompatible truths in Geometry, where the human mind prides itself on being able to see more clearly than elsewhere, all the more should I have cause to submit to the truths of an order beyond my reason, and remind myself always that He who created my reason was not obliged to make it capable of everything'].

[20] In *T*/30 (to show that existence belongs to units – twenty men can only exist because each of them exists).

puisse être, de vouloir concilier des vérités, dont le Createur a voulu lui cacher la compatibilité.[21]

Malezieu, a follower of Arnauld, shares this conviction with both Pierre Bayle and the Jesuits of the *Mémoires de Trévoux*. This is how Bayle puts the matter:

J'aurois beaucoup de choses à dire sur les Véritez Géométriques; mais je me contente de vous faire souvenir d'une chose dont Mr *Bernard* a parlé, en faisant mention des Elémens de Géometrie de *Mr. Le Duc* DE BOURGOGNE. C'est que ce sont deux Véritez également démontrées: l'une, que l'étenduë est composée de Parties infinies; l'autre, qu'elle ne l'est point. Les Jésuits des Trévoux, dans l'Extrait de cet Ouvrage du Duc DE BOURGOGNE, ont très-bien dévelopé cela, et en ont tiré une conséquence destinée à mortifier notre Raison, et à la disposer à croire les Mysteres les plus incompatibles selon notre Notions, à des Vérités evidentes. S'il y a Démonstration que les Parties d'un Pied de Matiere sont en nombre fini, il n'y a personne qui puisse se fier aux Démonstrations, par lequelles on prouve que la Ligne diagonale d'un Quarré contient une infinité de Parties; car pourquoi se feroit-on plûtot à ces Démonstrations-là qu'à celles du contraire? et ainsi vous voiez que rien n'est plus incertain, absolument parlant, que la Science Géométrique.[22]

This kind of scepticism, so easily associated with the irrational acceptance of faith, is evidently very distasteful to Hume – as he writes in the *Abstract*,

21 *Ibid.*, 'Réflection sur les incommensurables', Book IX, pp. 149–50 ['Thus is our reason reduced to strange extremities. Geometry proves the infinite divisibility of matter to us, and at the same time we find that it is composed of indivisibles. Let us humiliate ourselves once more, and recognise that it is not for a creature, however excellent it might be, to want to reconcile truths whose compatibility the Creator wanted to hide from it'].

22 Letter to Des Maizeau, 16 October 1705, *Oeuvres diverses de Mr. Pierre Bayle . . . contenant tout ce que cet Auteur a publié sur des matieres de Theologie, de Philosophie, de Critique, d'Histoire, et de Literature; excepté son Dictionnaire Historique et Critique*, 4 vols., The Hague 1727–31 (reprint Hildesheim 1964–70), vol. 4, Lettre CCCXXIV, p. 861 ['I should have many things to say about geometrical truths; but I shall content myself with drawing to your attention one thing Mr Bernard has talked about while mentioning the Elémens de Géometrie de Mr le Duc de Bourgogne. That is, that there are two truths equally proved: the one, that extension is composed of infinite parts; the other, that it is not so at all. The Jesuits of Trévoux, in the abstract of this work of the Duc de Bourgogne, have developed this very well, and have drawn from it a consequence destined to mortify our reason, and to dispose it to believe mysteries that are most incompatible according to our notions with evident truths. If there is a demonstration that the parts of a foot of matter are finite in number, nobody could trust the demonstrations through which one proves that the diagonal of a square contains an infinite number of parts; for why should one rely on these demonstrations rather than on those proving the contrary? And so you see that nothing is more uncertain, absolutely speaking, than geometrical science']. Jacques Bernard's review of Malezieu's *Elemens* is in *Les Nouvelles de la Republique des Lettres*, September 1705, pp. 340–57; the one in the *Memoires de Trévoux* is to be found in the issue of September 1705, Article 129, pp. 1467–79.

> 'Twere certainly to be wish'd, that some expedient were fallen upon to reconcile philosophy and common sense, which with regard to the question of infinite divisibility have wag'd most cruel wars with each other. (*A*/659)

So geometry does not appear to afford a very good model of knowledge; but, as I shall show, the *Treatise* does offer an alternative – of sorts. In order to do so, let me now start with a review of what is to be found, in the *Treatise*, about demonstrative knowledge and its main features.

DEMONSTRATIVE KNOWLEDGE

In the *Treatise*, one has to gather Hume's view of demonstrative knowledge from scattered observations.[23] For a start, it is worth reading again through the page which introduces the distinction between knowledge and probability together with the geometrical illustration of knowledge:

> There are seven different kinds of philosophical relation, *viz. resemblance, identity, relations of time and place, proportion in quantity or number, degrees in any quality, contrariety, and causation*. These relations may be divided into two classes; into such as depend entirely on the ideas, which we compare together, and such as may be chang'd without any change in the ideas. 'Tis from the idea of a triangle, that we discover the relation of equality, which its three angles bear to two right ones; and this relation is invariable, as long as our idea remains the same. (*T*/69)

While the three relations of identity, situation in space and time, and causation, are contingent associations dependent on experience, the four relations of resemblance, contrariety, degrees in quality, and proportion in quantity and number, depend solely on the related ideas, and constitute the 'objects of knowledge and certainty'. Of the four 'philosophical' relations constituting knowledge, three – resemblance, contrariety, and degrees in quality – are discovered 'at first sight, and fall more properly under the province of intuition than demonstration'. The fourth, proportion in quantity or number, is discovered by 'abstract reasoning and reflection'. Proportions in quantity or number allow us 'to carry on a chain of reasoning to any degree of intricacy', thus founding demonstrative knowledge (*T*/70): with them 'the mind can go beyond what is immediately present to the senses' to discover 'the relations of objects' (*T*/73). I shall show that in the *Treatise* this classic

[23] Observed in passing in Zabeeh, *Hume Precursor*, p. 113, in the introduction to his chapter on 'The domain of deductive reason': '. . . Hume's theory of deductive sciences, viz., arithmetic and geometry, which appears scattered throughout his work . . .'.

association of demonstrative and intuitive knowledge takes on a new and interesting meaning. But first I must consider Hume's notion of 'demonstration'.

In algebra and arithmetic 'we can carry on a chain of reasoning to any degree of intricacy, and yet preserve a perfect exactness and certainty' (*T*/71). These two branches of mathematics are 'certain and infallible sciences' because they transmit certainty from premises to their conclusions through a 'chain of reasoning'. So demonstrations are 'chains of reasoning', and there is the suggestion that they may be long and intricate. Moreover, we know that they differ from arguments based on the balancing of the evidence, because they are always either conclusive or no good at all:

> 'Tis not in demonstrations as in probabilities, that difficulties can take place, and one argument counter-ballance another, and diminish its authority. A demonstration, if just, admits of no opposite difficulty; and if not just, 'tis a mere sophism, and consequently can never be a difficulty. 'Tis either irresistible, or has no manner of force. (*T*/31)

A demonstration excludes the supposition of anything contradicting it as absolutely impossible:

> If such an inference [from the motion of one body from the impulse of another – effect from cause] may be drawn merely from the ideas of body, of motion, and of impulse, it must amount to a demonstration, and must imply the absolute impossibility of any contrary supposition. Every effect, then, beside the communication of motion, implies a formal contradiction; and 'tis impossible not only that it can exist, but also that it can be conceiv'd. (*T*/111)

And a contradiction is something which cannot either exist or be conceived. The link between contradiction and inconceivability is repeated again and again in the *Treatise*. For instance:

> 'Tis in vain to search for a contradiction in any thing that is conceiv'd by the mind. Did it imply any contradiction, 'tis impossible it cou'd ever be conceiv'd; (*T*/43)

> Such an inference [from cause to effect] wou'd amount to knowledge, and wou'd imply the absolute contradiction and impossibility of conceiving any thing different. (*T*/87)

> Whatever is absurd is unintelligible; nor is it possible for the imagination to conceive any thing contrary to a demonstration. (T/95)

That no demonstrative argument can be produced against something conceivable is likewise repeatedly stated; for instance:

Here then is an idea of extension, which consists of parts or inferior ideas, that are perfectly indivisible: consequently this idea implies no contradiction . . . these demonstrations cannot be just without proving the impossibility of mathematical points; which 'tis an evident absurdity to pretend to; (*T*/32–3)

But can any thing be imagin'd more absurd and contradictory than this reasoning? Whatever can be conceiv'd by a clear and distinct idea necessarily implies the possibility of existence; and he who pretends to prove the impossibility of its existence by an argument deriv'd from the clear idea, in reality asserts, that we have no clear idea of it, because we have a clear idea. 'Tis in vain to search for a contradiction in any thing that is distinctly conceiv'd by the mind. Did it imply any contradiction, 'tis impossible it cou'd ever be conceiv'd; (*T*/43)

The separation, therefore, of the idea of a cause from that of a beginning of existence, is plainly possible for the imagination; and consequently the actual separation of these objects is so far possible, that it implies no contradiction or absurdity; and is therefore incapable of being refuted by any reasoning from mere ideas; without which 'tis impossible to demonstrate the necessity of a cause. (*T*/79–80)

I have already observed that Hume's use of 'absurd' and 'contradictory' has a very special, very Humean flavour – certainly their sense has little to do with logic.[24] Let me give one further illustration of this (the passage is already familiar):

In common life 'tis established as a maxim, that the streightest way is always the shorter; which wou'd be as absurd as to say, the shortest way is always the shortest, if our idea of a right line was not different from that of the shortest way betwixt two points. (*T*/50)

That the shortest way is always the shortest is an obviously analytic proposition, not an absurdity; Hume's point is that the absurdity lies in saying it. This, in my opinion, must be taken very seriously indeed.

HUME'S A PRIORI BETWEEN KANT AND HUSSERL

I have hinted earlier that there is something interesting in Hume's version of the link between the certainty of a demonstrative argument and the confidence inspired by the immediate perception of, say, the resemblance between two objects. Let me consider for a moment what is going on in the page quoted above, in which Hume introduces the

[24] See above, chapter 1, fn. 46 and fn. 50. Cf. Gotterbarn, 'Kant, Hume and analyticity', and Suchting, 'Hume on necessary truth'.

seven philosophical relations and, with them, the distinctions between knowledge and probability, and between intuitive and demonstrative knowledge. The focus of Hume's argument is obviously on probability; knowledge is only introduced as the foil of probability. Its main feature, from this point of view, seems to be its lack of contact with outside perception: the four relations of resemblance, contrariety, degrees in quality, and proportions of quantity or number, the *Treatise* says, 'depend entirely on the ideas, which we compare together' (*T*/69), that is, they are independent of experience, and 'depending solely upon ideas, can be the objects of knowledge and certainty' (*T*/70). In other words, these relations are only established, in fact only exist, as operations of the mind. Moreover, their certainty is the outcome of certain acts of the mind, a feature of certain mental processes; it lies in the way the mind works. They are, one may say, a priori, even though, once more, not exactly in Kant's sense. This point deserves some further clarification – and I shall offer it by discussing briefly the readings of Hume offered by Husserl and his students.

Phenomenologists have regarded Hume's notion of the a priori as a major testimony to his philosophical greatness because, as Husserl puts it, for the first time it is an a priori which is immediately given in the act of consciousness, and is therefore intuitively self-evident.[25] This a priori appears in the *Treatise* in two guises.[26] The first is, as I have said above, the theory of the relations of ideas. The second manifestation of the a priori has to do with a process which makes its first appearance in the discussion of abstract ideas.

Kant's attention is focussed on Hume's discussion of the a priori in mathematics, that is, of the proportions of quantity and number. On the contrary, phenomenologist readers tend to concentrate on the a priori as it is involved in the other 'philosophical' relations – for, as Reinach puts it, 'what is a plausible thesis with regard to mathematical propositions is, with regard to the other relations of ideas, a plain absurdity'. To uncover the general mechanism of association operating in the relation of ideas, Reinach chooses to concentrate on resemblance, as it appears, for example, in the proposition 'Red and Orange are similar'.[27] Of course, according to Hume, our ideas of red and

[25] See Murphy, *Hume and Husserl*, chapter 2, 'The a priori and evidence', esp. p. 33.

[26] See Livingston, *Hume's Philosophy*, pp. 48ff.

[27] Reinach, 'Kant's interpretation', p. 168. The similarity between two colours appears in the *Treatise*, but curiously Reinach has changed the colours!: '*Blue* and *green* are different simple ideas, but are more resembling than *blue* and *scarlet*; tho' their perfect simplicity excludes all possibility of separation or distinction' (*T*/637). (On this passage see below, chapter 4,

orange must come from experience – they are copied from impressions. Yet, once the ideas have been acquired, the proposition stating the resemblance between them is clearly a priori: for the relation 'depends entirely on the ideas, which we compare together', that is, it is grounded in the nature of the two colours; its contrary is impossible; and 'knowledge of it is possible, quite independently from experience, through the mere representations of the two colours' (p. 168). But it is also obviously not analytic. Hume's a priori suddenly appears much more interesting.

And it becomes even more interesting if one considers its other side, the one having to do with abstract ideas. Hume wrote that 'abstract ideas are in themselves individual, however they may become general in their representation' (*T*/20). This is what another of Husserl's pupils, Christopher Salmon, makes of it:

> we should look for the origin of abstract ideas in our actual abstraction of them. If we can never 'perceive' anything abstract, and so cannot assert that anything abstract 'really exists', we can, perhaps, create abstract ideas for ourselves by a certain process of the 'imagination'. Hume's 'however they may become general in their representation' makes the reader curious to know how that could happen, and suggests to him that Hume may have had some notion as to how it actually does happen. And indeed Hume had. Having denied the 'existence' of abstract ideas, he offers a very tolerable description of how they are abstracted.[28]

And this description (in *T*/20–2) is, Salmon argues,

> the first practical instance in the *Treatise*, of Hume's theory that the ultimate 'explanation' of our consciousness of objects lies, not in the objects, but in the processes of consciousness itself. We can only be conscious of objects the consciousness of which we have constructed for ourselves. (p. 336)

According to Salmon, the process of abstraction involved in this activity is, in turn, a form of introspective practice. And had he pursued it consistently, instead of falling back into a Lockean associative psychology, Hume would have gone in the direction of exhausting 'the types not only of all actual, but also of all possible, experiences of objects' (p. 323). In short, he would have been the first major phenomenologist.

The phenomenological approach has attracted the attention of

'Simplicity and distance'.) On resemblance see also R. W. Church, *Hume's Theory of the Understanding*, London 1968.

[28] C. Salmon, 'The central problem of David Hume's philolosphy: an essay toward a phenomenological interpretation of the first book of the *Treatise of Human Nature*', *Jahrbuch für Philosophie und phänomenologische Forschung*, x, 1929, pp. 299–449, esp. p. 334.

some recent Anglo-American readers of Hume. In particular, Donald Livingston's reaction to it has produced very interesting results. According to Livingston, Humean 'perceptions' are entities that are described only with difficulty in the eighteenth-century Lockean language available to Hume. He was therefore forced to present them 'narratively':

> Perhaps enough has been said to show that the Locke–Berkeley–Hume story obscures rather than illuminates Hume's concept of 'perceptions'. Freeing ourselves from it, and appoaching the *Treatise* as a piece of literature, we find that the concept of 'perception' is internal to a narrative order of exposition. The subject of a narration becomes more intelligible as the story unfolds, and the reader is required to view earlier episodes in the light of later episodes.[29]

In order to make sense of the nature of Humean perceptions, it is necessary to take into account the two ways in which the a priori makes its appearance in the *Treatise* – in the theory of relations of ideas and in the abstractive process examined by Salmon. (This latter is identifiable with the distinction of reason, the process through which our mind singles out such inseparable aspects of experience as, for example, the colour and the figure of a white globe when we realise its similarity on the one hand with a black globe, and on the other with a white cube.)[30] Each of these two ways, Livingston writes,

> frames a conception of necessity which enables us to make judgments that are neither analytic nor synthetic about the a priori structure of perceptions. (p. 49)

Livingston follows Reinach in rejecting the assimilation of the distinction between the relations depending solely on the ideas and the relations depending on experience to the Kantian distinction between the analytic and the synthetic (pp. 50ff.); and he follows Salmon in arguing 'against the received view that Hume's method of analysis is that of breaking experience down into phenomenal atoms' (pp. 53ff.). The reading that Livingston sketches as resulting from this interpretation of the Humean a priori is fascinating and convincing:

> we may explicate a perception . . . as (1) a propositional and intentional content

[29] Livingston, *Hume's Philosophy*, p. 49.

[30] This abstractive mechanism is discussed, with the example of the globe of black marble etc., in the *Treatise* at the end of the section on abstract ideas (*T*/24–5). The distinction of reason also accounts for the possibility of finding a resemblance between simple perceptions: see also *T*/637, another passage much loved by phenomenologists. See the fundamental essays by R. J. Butler, '*Distinctiones rationis*, or the Cheshire Cat which left its smile behind it', *Proceedings of the Aristotelian Society* N.S. 76, 1976, pp. 165–76, and 'Hume's impressions', in G. Vesey (ed.), *Impressions of Empiricism*, London 1976, pp. 122–36.

inseparably connected in experience to (2) an act of the mind (felt as force and vivacity) where (1) and (2) can be differentiated only through a distinction of reason. That is, the act of the mind that is the force and vivacity of a perception is connected to its propositional content in the same way in which figure and colour are connected in the perception of a blue circle. The whole perception, of course, is an act of the mind. The force and vivacity of a perception is simply that aspect, uncovered by a distinction of reason, that is '*felt* rather than conceiv'd' . . . The force and vivacity of any perception is our awareness of what its propositional and intentional content *means* to us. (pp. 57–8)

So Livingston underlines at once the wholeness of the perceptual act, and the way in which we come to see the content (that we conceive), and the impact of that content on us, its meaning-for-us (that we feel), as distinct aspects of that same experiential whole.

Let us consider the Humean a priori as it appears in one of the relations, that of contrariety, and in its link with the ideas of existence and non-existence. Being a relation, contrariety must imply some sort of resemblance between the ideas being associated – for without resemblance, we are told, no philosophical relation can exist, 'since no objects will admit of comparison, but what have some degree of resemblance' (*T*/14). After this premise, contrariety is introduced as follows:

the relation of *contrariety* may at first sight be regarded as an exception to the rule, *that no relation of any kind can subsist without some degree of resemblance*. But let us consider, that no two ideas are in themselves contrary, except those of existence and non-existence, which are plainly resembling, as implying both of them the idea of the object. (*T*/15)

So resemblance and contrariety are, for opposite reasons, a bit special, a bit different from the other relations of ideas – resemblance, because it always holds whenever any other relation holds, and contrariety because it only holds between the two very special ideas of existence and non-existence. This relation is an object of immediate, intuitive knowledge:

no one can once doubt but existence and non-existence destroy each other, and are perfectly incompatible and contrary. (*T*/70)

This definition is put to use later to state that 'any thing may produce any thing' (*T*/173). As Hume puts it, this makes very good sense

if we compare the two principles explain'd above, *that the constant conjunction of objects determines their causation*, and *that properly speaking, no objects are contrary to each other, but existence and non-existence*. Where objects are not contrary, nothing hinders them from having that constant conjunction, on which the relation of cause and effect totally depends. (*T*/173)[31]

The idea of existence is not, Hume says, a separate and distinct idea.[32] It coincides with the idea of the object that is a candidate for existence.

The idea of existence . . . is the very same with the idea of what we conceive to be existent. To reflect on any thing simply, and to reflect on it as existent, are nothing different from each other. That idea, when conjoin'd with the idea of any object, makes no addition to it . . . Any idea we please to form is the idea of a being; and the idea of a being is any idea we please to form. (*T*/66–7)

Now the first problem is, in what sense can existence and non-existence be said to be contrary to each other? For, as Hume himself seems to realise, there is an immediate difficulty here – let us read his definition again, this time in its entirety:

no two ideas are in themselves contrary, except those of existence and non-existence, which are plainly resembling, as implying both of them an idea of the object; tho' the latter excludes the object from all times and places, in which it is supposed to exist. (*T*/15)

So the difference between an existent angel Gabriel and a non-existent one turns out, on this account, to be that the first exists say in Heaven, while the second fails to exist in Heaven. As Passmore sums it up:

To assert the non-existence of anything, it now appears, is to say that it does not occur at certain places and times; to assert its existence is to say that it does occur at those places and times.[33]

The point is, in his opinion, that Hume's position that all we conceive is conceived as existent is open to the typical Eleatic objection – if this is so, non-existence is inconceivable. In order to avoid the objection, Hume falls back on the conception of non-existence at a certain place and time. But it is evident, according to Passmore, that something has gone wrong with Hume's theory, since it cannot sort out the problem of 'what an "idea of non-existence" could be' (p. 26) without '[violently departing] from his general presuppositions', and which, more generally, 'cannot account for affirmation and negation' (p. 28). It is simply

31 Similarly, in *T*/247 that 'no real objects are contrary' because contrariety only holds between existence and non-existence is used to show that there cannot be a priori arguments against thinking matter.

32 Cf. Kant's 'existence is not a predicate'.

33 *Hume's Intentions*, p. 26. On the same passages see also D. Pears, *Hume's System: An Examination of the First Book of His 'Treatise'*, Oxford 1990, pp. 55–9.

not true that 'whatever we conceive, we conceive to be existent' (*T*/67). And Passmore refers to a passage in the *Enquiry*:

That Caesar, or the angel Gabriel, or any being never existed may be a false proposition, but still is perfectly conceivable, and implies no contradiction. (*E*/164)

He comments that 'if to conceive a thing is to conceive it as existing, then it *would* "imply a contradiction" to conceive the non-existence of Gabriel'; 'if we cannot think of a thing except as existing, as Hume has previously argued, then we obviously cannot think of it as non-existent; and to contemplate "the negation of a fact" would involve thinking simultaneously of an idea as existing (the fact) and as non-existing (the negation)'.[34]

Passmore's analytic reading seems to be fair and inspired by great common sense; it is however an analytically informed sort of common sense. A first inkling as to how the phenomenologist may avoid the impasse he points out comes from Salmon, who argues that, according to Locke's theory of knowledge as conformity, one would become conscious of, say,

the real house when the last part of the ideal house, a window or a chimney-pot perhaps, had been stuck into its place in [one's] understanding.[35]

Establishing to what extent this is a fair comment on Locke is beyond the scope of my discussion. What I find interesting is that this certainly does not apply to Humean perceptions. For, in the *Treatise*, while introducing the so-called copy principle – that all simple ideas make their first appearance to the mind in corresponding impressions – Hume notes:

many of our complex ideas never had impressions, that corresponded to them, and . . . many of our complex impressions never are exactly copied in ideas. I can imagine to myself such a city as the *New Jerusalem*, whose pavement is gold and walls are rubies, tho' I never saw any such. I have seen *Paris*; but shall I affirm that I can form such an idea of that city, as will perfectly represent all its streets and houses in their real and just proportions? (*T*/3)

This passage starts by talking of an imagined existence which does not involve any actual, real exemplification of the 'existent' object in the outside world, an 'existence' having to do with the lively evocation of

34 Passmore, *Hume's Intentions*, pp. 98 and 27. According to Passmore, p. 98, Hume 'is left with no way of distinguishing between "having the *form* of a fact" and "being a fact"'.
35 *Ibid.*, p. 307.

ideas – in the passage quoted above, the New Jerusalem has a more convincing tinge and feels more, as it were, 'existent' than Paris, on account of those pavements of gold and walls of rubies which inevitably capture our imagination far more than the flat reference to the 'streets and houses' in Paris. And the similarity between the non-existent New Jerusalem and the existent Paris is in the working of the imagination, in both cases involved in the construction of the idea. This is, I think, the key to make sense of the case of the existence or non-existence of the angel Gabriel: the conception of the place in which the non-existent angel fails to exist is only triggered (but then it is triggered necessarily) at the level of the ideas of things; while the contrariety between the ideas of existence and non-existence only applies at the level of the ideas of mental operations and of perceptions of things.

So contrariety only relates the acts of the mind involved in the conception of ideas. Similarly, when Hume says that there is no difference between conceiving something, and conceiving it as existent, he is clearly pointing to the fact that existence is not, as such, an independent mental content added to that particular something that we are conceiving as existent. For the idea of existence is the idea of a manner of conception, a particular way in which that something, that particular mental content, is conceived. Hume also establishes a relation between the idea of existence and another, particularly important 'manner of conception' which I have examined earlier – belief. In his classic analytic reading of the British empiricists, Jonathan Bennett is very disapproving of this move of Hume's: this 'supposed connexion' of the idea of existence with belief is 'a bad mistake', he writes.[36] But then Bennett disapproves of the whole account of belief to be found in the *Treatise*:

> [Hume] thinks that the common propositional content which a belief, disbelief, etc. can share is an idea, which implies that believing does not differ from disbelieving or wondering, say, in the ideas that are involved. In inferring from this that the difference can only consist in 'the manner in which' the ideas are had, Hume is assuming that someone's having a certain belief is just a fact about what ideas he has and how he has them. This assumption is unwarranted, and the resultant analysis of belief is beyond redemption. It implies, for example, that there is no difference between believing that the Sahara is warm and entertaining the thought that it is extremely hot! (p. 294)

Bennett is certainly right – Hume's account of belief fails to adequately rule out the transference of vivacity between content and manner of

[36] J. Bennett, *Locke, Berkeley, Hume: Central Themes*, Oxford 1971, pp. 294–5, fn. Bennett refers to P. T. Geach, 'Assertion', *Philosophical Review* 74, 1965, pp. 458–9.

conception. And, more generally, in so far as Hume's talk about belief is an analysis, it really does appear to be beyond redemption.

And yet Bennett's crisp arguments miss something interesting in Hume's treatment of belief. Of course, Hume is aware of the problem pointed out by Bennett. He also suggests in what direction we are to look for a solution:

We observe, that the vigour of conception, which fictions receive from poetry and eloquence, is a circumstance merely accidental, of which every idea is equally susceptible; and that such fictions are connected with nothing that is real. This observation makes us only lend ourselves, so to speak, to the fiction: But causes the idea to feel very differently from the eternal establish'd persuasions founded on memory and custom. They are somewhat of the same kind: But the one is much inferior to the other, both in its causes and effects. A like reflexion on *general rules* keeps us from augmenting our belief upon every increase of the force and vivacity of our ideas. Where an opinion admits of no doubt, or opposite probability, we attribute to it a full conviction; tho' the want of resemblance, or contiguity, may render its force inferior to that of other opinions. 'Tis thus the understanding corrects the appearances of the senses, and makes us imagine, that an object at twenty foot distance seems even to the eye as large as one of the same dimensions at ten. (*T*/632)

So Hume states a distinction between different kinds of vivacity and, even though he does not justify it, let alone fully develop it, he does indicate the presence of a family of operations which compensate for the vagaries of accidental experience.[37]

To what Hume calls the 'unphilosophical variety of terms' used in the *Treatise* to describe belief there corresponds, I think, an 'unphilosophical variety' of ways in which an idea may be strong and lively – so that, in Hume's terms, it is still possible to see the difference between the entertained idea of something lively (a hot Sahara or the New Jerusalem), and the belief in the existence out there of something dull (a warm Sahara or Paris). Consider what Hume writes about the difference between reading history and reading romance:

A poetical description may have a more sensible effect on the fancy, than an historical narration. It may collect more of those circumstances, that form a

[37] On the different kinds of vivacity and on general rules in the *Treatise*, see J. Laird, *Hume's Philosophy of Human Nature*, London 1932, pp. 91–2; Passmore, *Hume's Intentions*, pp. 10, 63–4 and passim; A.-L. Leroy, *David Hume*, Paris 1953, pp. 69ff.; T. K. Hearn, Jr., '"General rules" in Hume's *Treatise*', *Journal of the History of Philosophy* 8, 1970, pp. 405–22; W. H. Walsh, 'Hume's concept of truth', in G. Vesey (ed.), *Reason and Reality*, London 1972, pp. 99–116; and my own 'Quixotic confusions and Hume's imagination: the "science of human nature" on the beliefs and sentiments of readers' (forthcoming).

compleat image or picture. It may seem to set the object before us in more lively colours. But still the ideas it presents are different to the *feeling* from those, which arise from the memory and judgment. There is something weak and imperfect amidst all that seeming vehemence of thought and sentiment, which attends the fictions of poetry. (*T*/631)[38]

Here the point of view is reversed – the emphasis is all on the difference, and it is something original, a difference in sentiment. And it is worth remembering also, in this connection, that the relation of contrariety appears again in Book 2 of the *Treatise*: where one finds that contrariety holds between opposite passions, for example between love and hatred, in so far as they have the same object. The point Hume is making is that the object of these passions is not to be identified with their cause:

But tho' the object of love and hatred be always some other person, 'tis plain that the object is not properly speaking, the *cause* of these passions, or alone sufficient to excite them. For since love and hatred are directly contrary in their sensation, and have the same object in common, if that object were also their cause, it wou'd produce these opposite passions in an equal degree; and as they must, from the very first moment, destroy each other, none of them wou'd ever be able to make its appearance. (*T*/330)[39]

Here there is the idea of an operation of the mind interrelating two other natural operations, this time of our passions, and in this case it is no surprise that the liveliness belongs, unambiguously, to the 'manner of conception'.[40]

At this point I think it is clear what the relation of contrariety shows about the Humean a priori. This relation only seems to make sense if one interprets it as presupposing a distinction, within the wholeness of the perceptual act, between the content and the 'manner' in which this is perceived. And the ideas of existence and non-existence are related to the whole family of ideas of 'manners', of appearance and of conception, one of which, the idea of space, is the subject of the present study. The feature that distinguishes them all is beyond their content – like the difference in the feelings of an eighteenth-century polite reader when reading a history book rather than a novel, or the difference between the emotional colour of an object of passion when it inspires a feeling of love rather than of hatred. We may conclude that they all belong to our

38 See also *T*/121ff.

39 The same is to be found about pride and humility (*T*/278). In *T*/441 contrary passions are said to occur alternately when they are aroused by different objects; and in *T*/ 464 the list of the relations depending solely on ideas appears again in the argument showing that morality does not depend on demonstrative reason.

40 See Butler, 'Cheshire Cat' and 'Hume's impressions'.

emotional life – which is also where one finds the Humean a priori. But now it is time to go back to mathematical knowledge and to the sense in which it can be said to be certain.

A PASSION FOR TRUTH

So the certainty discussed in the *Treatise* lies in the way the mind works: it exists as a feature of a particular kind of acts of the mind – it does not lie where one would have thought it did. This is similar to the case of the idea of necessary connection between a cause and its effect, and Hume himself actually puts forward such a comparison:

> as the necessity, which makes two times two equal to four, or three angles of a triangle equal to two right ones, lies only in the act of the understanding, by which we conceive and compare these ideas; in like manner the necessity or power which unites causes and effects, lies in the determination of the mind to pass from the one to the other. (*T*/166)[41]

Cassirer aptly comments on this passage:

> when we speak of cause and effect, of force and necessity, all these expressions denote nothing which was to be found in the objects themselves; rather, they first gain their sense in the *consideration* of the spirit.[42]

The idea of certainty originates from an impression of reflection – it is the appearance in thought, as it were, of how our own mind feels to us when it is at work in some particular ways. These particular ways are not specific to demonstrative knowledge, which shares 'certainty' with other forms of knowledge – for example, with the idea of cause and effect, that is, with areas of probability. More seriously, the very possibility of 'certainty' as it is commonly meant in relation to demonstrative knowledge seems to be seriously shaken. Can one conclude that, in principle, there is no space for demonstrative knowledge and its certainty in the universe of the *Treatise* – that is, that Hume only left it in because he was, as it were, absent-minded? I do not think so. I would put it as follows. In Hume's text there is an unresolved tension between different ways of conceiving certainty. The usage of such terms as 'demonstration', 'contradiction', 'contradictory', and 'absurd' in his pages suggests that, when it is made the object of a science of human

[41] Cf. above, chapter 2, 'A philosopher's eye' (esp. fn. 17).

[42] 'Wenn wir von Ursache und Wirkung, von Kraft und Notwendigkeit sprechen, so bezeichnen all diese Audrücke Nichts, was sich an dem Gegenständen selbst vorfände, sondern sie gewinnen erst in der *Betrachtung* des Geistes ihren Sinn' (*Erkenntnisproblem*, p. 260).

nature, the certainty of mathematics tends to be assimilated to a feeling. On the other hand, it is evident that demonstrative knowledge, in a more standard sense, also exists and has a part to play in the *Treatise*. So the two dimensions coexist: the opposite of a demonstration is called both impossible and inconceivable, and the certainty discussed in the *Treatise* involves both analytic rigour, and persuasion and assent. The ambiguity of Hume's treatment of certainty, it seems to me, results in a shift of emphasis, from a traditional and rigorous notion of certainty to a less crisp and more comprehensive one which has to do both with whether something, for instance mathematics, is certain, and with how certain we feel about something, for instance a mathematical demonstration. With this in mind we can move, now, to consider 'truth' and the passional dimension of the search after it.

To what extent and by what means does a demonstrative argument command our assent? Or, in other words, what is the relation between demonstrative certainty and persuasiveness? In the discussion of belief, there is a splendid description of the confidence inspired by intuitive and demonstrative knowledge:

> Wherein consists the difference betwixt believing and disbelieving any proposition? The answer is easy with regard to propositions, that are prov'd by intuition or demonstration. In that case, the person, who assents, not only conceives the ideas according to the proposition, but is necessarily determin'd to conceive them in that particular manner, either immediately or by the interposition of other ideas. (*T*/95)

Here, knowledge – intuitive and demonstrative – seems to compel the mind to an immediate assent. But things may not be quite so easy and straightforward:

> as none will maintain, that our assurance in a long numeration exceeds probability, I may safely affirm, that there scarce is any proposition concerning numbers, of which we have a fuller security. For 'tis easily possible, by gradually diminishing the numbers, to reduce the longest series of addition to the most simple question, which can be form'd, to an addition of two single numbers; and upon this supposition we shall find it impracticable to shew the precise limits of knowledge and probability, or discover that particular number, at which one ends and the other begins. But knowledge and probability are such contrary and disagreeing natures, that they cannot well run insensibly into each other, and that because they will not divide, but must be either entirely present, or entirely absent. Besides, if any single addition were certain, every one wou'd be so, and consequently the whole or total sum; unless the whole can be different from all its parts. I had almost said, that this was certain; but I

reflect, that it must reduce itself, as well as any other reasoning, and from knowledge degenerate into probability. (*T*/181)

It is evident that this is Hume in his most radical and disruptive mood; and yet I find that here there is something very important and very different from a merely Pyrrhonian exercise. Let me put it as follows. The certainty of a demonstration and the assent its result actually commands may occasionally coincide; but our confidence in performing the first steps, at the beginning of the argument, is reduced a bit at each argumentative step, no matter how elementary and safe, and for long and intricate arguments the loss of confidence may be considerable.[43] In other words, a demonstrative argument may be certain, and yet we may not be certain about it. On the other hand, the contrary of a demonstration should, in principle, be inconceivable; but obviously in reality this may not be the case, for if the result is far-fetched enough the mind will find no difficulty in conceiving the opposite – is it true, to take an example from the *Treatise* itself, that 'the angles of a chiliagon [are] equal to 1996 right angles' (*T*/72)? In short, the emphasis in Hume's pages is not on the definition of demonstrative knowledge or on the understanding of what demonstrative knowledge is like, but rather on the process of demonstrative knowing and its relation to assent and to how certainty feels to us. Mathematics and demonstrative knowledge are in human hands, and

there is no Algebraist nor Mathematician so expert in his science, as to place entire confidence in any truth immediately upon his discovery of it, or regard it as any thing, but a mere probability. Every time he runs over his proofs, his confidence encreases: but still more by the approbation of his friends; and is rais'd to its utmost perfection by the universal assent and applauses of the learned world. (*T*/180)

This means that even the most absolute necessity of any demonstrative argument is limited, at the level of the assent commanded, by the human dependence on the repetition of proofs and on its acceptance first by friends and then by the learned community: from this point of view, truly, 'demonstration is subject to the control of probability' (*T*/182).

So even with regard to mathematics, in fact especially with regard to mathematics,

our reason must be consider'd as a kind of cause, of which truth is the natural effect. (*T*/180)

[43] For stimulating comments on this see R. Fogelin, *Hume's Skepticism*, chapter 2, 'Hume's skepticism concerning reason', pp. 13–24.

The other side of this coin appears in Book 2 of the *Treatise*, where the whole concluding section is dedicated to 'curiosity, or the love of truth' – 'that love of truth, which was the first source of all our enquiries' (*T*/448), as Hume presents it.[44] It is a passion with a particular status:

'tis an affection of so peculiar a kind, that 'twould have been impossible to have treated of it under any of those heads, which we have examin'd, without danger of obscurity and confusion.

Mathematics appears again, and in the first position; in fact, it occupies four of the six pages of this section. The 'species of truth' consisting in 'the discovery of the proportions of ideas, consider'd as such'

is not desir'd merely as truth, and . . . 'tis not the justness of our conclusions, which alone gives the pleasure. For these conclusions are equally just, when we discover the equality of two bodies by a pair of compasses as when we learn it by a mathematical demonstration; and tho' in the one case the proofs be demonstrative, and in the other only sensible, yet generally speaking, the mind acquiesces with equal assurance in the one as in the other. (*T*/449)

There are two elements in the pleasure of discovering truth demonstratively: the exciting game of the operations of the understanding, and the significance of the truth discovered. The latter is evident from the fact that

'tis easy to multiply algebraical problems to infinity, nor is there any end in the discovery of the proportions of conic sections; tho' few mathematicians take any pleasure in these researches, but turn their attention to what is more useful and important. (*T*/449–50)

It is the first, however, that really matters:

The first and most considerable circumstance requisite to render truth agreeable, is the genius and capacity, which is employ'd in its invention and discovery. (*T*/449)

Human nature rules here. But the lovers of truth are strange creatures, and the two elements of the love of truth cannot be easily separated:

many philosophers have consum'd their time, have destroy'd their health, and neglected their fortune, in the search of such truths, as they esteem'd important and useful to the world, tho' it appear'd from their whole conduct and behaviour, that they were not endow'd with any share of public spirit, nor had any concern for the interests of mankind. Were they convinc'd, that their discoveries were of no consequence, they wou'd entirely lose all relish for their

[44] It is interesting that here, while writing the *Treatise*, Hume mentions love of truth, rather than of literary fame, as the first source of his work.

studies, and that tho' the consequences be entirely indifferent to them; which seems to be a contradiction. (*T*/450)

The misanthropic philosopher who has sacrificed his whole life to the search after a useful truth is no different from other men dominated by a 'ruling passion'. 'There cannot be two passions more nearly resembling each other, than those of hunting and philosophy' (*T*/451): as in philosophising, in hunting the pleasure is in 'the action of the body and mind; the motion, the attention, the difficulty, and the uncertainty'; in both cases 'these actions must be attended with an idea of utility, in order to their having an effect upon us', and in both cases

the end of our action may in itself be despis'd, yet in the heat of the action we acquire such an attention to this end, that we are very uneasy under any disappointments, and are sorry when we either miss our game, or fall into any error in our reasoning. (*T*/452)

Truth is the philosophical quarry. And this is not all – it is also what philosophers gamble on:

If we want another parallel to these affections, we may consider the passion of gaming, which affords a pleasure from the same principles as hunting and philosophy. (*T*/452)

Finally, the love of truth which inspires philosophy also has a less aristocratic relative – another disinterested pursuit:

besides the love of knowledge, which displays itself in the sciences, there is a certain curiosity implanted in human nature . . . Some people have an insatiable desire of knowing the actions and circumstances of their neighbours, tho' their interest be no way concern'd with them. (*T*/453)

So much for the magnificent pretensions of the philosophers and their disciples: philosophising is like hunting and gambling, and has a close relation to gossip – and reason itself is, in the end, nothing but a natural instinct.

CHAPTER 4

Talking about a vacuum: The paradoxes of empty space in section 5

> Every thing which Hume has advanced against Christianity had passed through my mind long before he wrote. Always remember this, that after a system is well settled upon positive evidence, a few partial objections ought not to shake it. The human mind is so limited, that it cannot take in all the parts of a subject, so that there may be objections raised against anything. There are objections against a *plenum*, and objections against a *vacuum*; yet one of them must certainly be true.
>
> (Samuel Johnson, in J. Boswell, *The Life of Samuel Johnson*, Thursday 21 July 1763, Oxford 1980, p. 314)

> Hopeless tasks are not always thankless.
>
> (Stanley Cavell, 'Existentialism and analytical philosophy', in *Themes out of School: Effects and Causes*, Chicago 1984, p. 203)

INTRODUCTION

The preceding chapters have shown that the idea of space as treated by Hume is a very paradoxical item. Space is conceived as composed of unextended coloured and tangible points – points, one is explicitly told, which are to be regarded as neither mathematical nor physical. In theory, each point is a distinct impression, each impression of a point a distinct perception – but, in practice, the points are much too small and too numerous for us to operate with them at all. This does not mean, however, that one is to conceive them as either infinitely small or infinitely numerous. Moreover, the idea of space also depends on the 'manner of appearance' of the points, that is to say on something about which, as I have shown, at least one thing is clear: it is not easily assimilated into any such simple position as that implied by Hume's principle of the correspondence between impressions and ideas – the very principle he keeps citing as specially relevant to the case of the idea of space. So Hume's treatment of the idea of space is at once atomistic and relational, and both strictly empirical and not-entirely-sensational. All this creates a family of problems linked with each other in complex, indirect ways. But I hope my reader will concur that this family of problems has produced no terrible impasse so far, and that it has been possible to deal with it in a reasonably reassuring way – it is only in the section Hume devotes to the discussion of the idea of a vacuum that it gives rise to a true crisis.

The treatment of the idea of empty space in Part 2, section 5 is, in fact, one of the most difficult parts of the whole *Treatise*. A few lines from this section will be enough to make this clear:

> I shall conclude this subject of extension with a paradox, which will easily be explain'd from the foregoing reasoning. The paradox is, that if you are pleas'd to give to the indivisible and intangible distance, or in other words, to the capacity of becoming a visible and tangible distance, the name of a vacuum, extension and matter are the same, and yet there is a vacuum. If you will not give it that name, motion is possible in a plenum, without any impulse *in infinitum*, without returning in a circle, and without penetration. But however we may express ourselves, we must always confess, that we have no idea of any real extension without filling it with sensible objects, and conceiving its parts as visible or tangible. (*T*/64)

'I am not sure that I understand what Hume is here intending to say', Kemp Smith begins the brave, contorted paragraph in which he tries to escape his interpretative predicament:

> If we have no idea of any extension which is not occupied by sensible objects, and if on this account we describe the non-sensible species of distance as being a vacuum, extension and matter are then being taken to be one and the same, and we must not mean by a vacuum anything extended . . . if non-sensible distance is not allowed to be a vacuum, and instead is assumed to be a plenum, then in view of the features, as above described, proper to non-sensible distance, motion is possible in what is 'without parts, without composition, invariable and indivisible', and which therefore does not allow either impulse or penetration . . .[1]

A valiant attempt – but not, I think, a successful one.

It is evident that making sense of this passage is much easier if, rather than trying a metaphysical interpretation, one chooses to place it in the context of the natural philosophical debates of the time. The coexistence of the identification of space and extension with the use of the

[1] N. Kemp Smith, The *Philosophy of David Hume*, London 1941, p. 316.

term 'vacuum' appears incomprehensible at the purely theoretical level. But with Flew one may easily make sense of it:

> It would seem to follow on Humean principles that when the physicists talk about vacua they must be talking nonsense, and that when they purport to demonstrate to us the evacuation of vessels these pretended vacua must really be plena. But these conclusions – both so embarrassing to anyone raised in Britain under the enormous shadow of Newton, and so apparently inconsistent with experience – are not drawn.[2]

What Flew says seems sensible, if a bit disappointing. A similar, but perhaps more sympathetic reading would presumably note that empty space was, in the writings of the Newtonians, the 'space of God'. So perhaps the identification of space and extension expresses Hume's resistance to this aspect of the Newtonian natural philosophical enterprise without rejecting Newtonianism altogether, and, in fact, while accepting and even trying to promote a different interpretation of it.[3] Both in Flew's interpretation and in this one I have sketched, there is a blatant contradiction in Hume's thought, which is explained contextually: in order to escape the dilemma of either opposing Newtonianism, or accepting its theological side, Hume produces a theoretical monstrosity. This is not all that serious, according to this interpretative line, since the discussion of empty space is, at any rate, very marginal in the *Treatise*.

I shall propose another interpretation, one which brings out the significance of this section and is both historically and metaphysically acceptable. After a summary of Hume's argument and its difficulties, I shall prepare the ground for this interpretation by reconstructing the terms of contemporary debates on empty space; in particular, I shall focus on Edmund Law's footnotes on the ideas of space and of a vacuum in his 1731 English translation of William King's *De origine mali*, I shall check the key-words of the debate in Chambers' *Cyclopaedia*, and I shall briefly consider relevant passages from texts by Descartes, Locke,

[2] A. Flew, 'Infinite divisibility in Hume's *Treatise*', in D. W. Livingston and J. T. King (eds), *Hume. A Re-evaluation*, New York 1976, p. 267. See also J. Wright, *The Sceptical Realism of David Hume*, Manchester 1983, p. 105: Hume's 'official reason for claiming that there is such inadequacy [in our ideas of space and time] is the existence of the opposing natural instinct. However, it is tempting to think that the real reason has more to do with the fact that Newtonianism was the real scientific philosophy becoming dominant in Hume's own day.' Cf. J. Noxon, *Hume's Philosophical Development*, Oxford 1973, p. 114: 'He denies the reality of absolute space, pointing out that in the absence of any visible or tangible thing there would be no recognition of space. Accordingly he identifies space and extension.'

[3] An example of this latter reading is to be found in my own 'Hume sullo spazio vuoto', *Studi Settecenteschi* (7–8), 1985–6, pp. 191–204.

and Barrow. Only at this point shall I start to press questions on the *Treatise* in earnest. My questions will concentrate on a few major points: the idea of space as distance and the relation of the notion of distance with that of space as 'manner of appearance', the evocative power of words, and the relation between different kinds of metaphysics.

A VERY DIFFICULT PIECE OF WRITING

It may be useful to begin with a survey of the arguments presented in the approximately twelve pages of section 5, where, under the unexciting title 'The same subject continu'd', Hume presents the discussion of the idea of empty space. I shall summarise it, for a start, with very little comment.

Given that '*the idea of space or extension is nothing but the idea of visible and tangible points distributed in a certain order*' – so the section begins – it is clear that the 'idea of a vacuum, or space, where there is nothing visible or tangible' is inconceivable (*T*/53). Therefore the standard arguments for the conceivability and existence of empty space turn out to be objections to Hume's theory. Hume examines three of them. First, 'the very dispute is decisive concerning the idea': the fact that we talk about empty space is in itself evidence that we have an idea of it (*T*/54). Secondly, the idea of a vacuum is 'a necessary and infallible consequence' of the 'two possible ideas of *rest* and *annihilation*':

> what must we conceive to follow upon the annihilation of all the air and subtle matter in the chamber, supposing the walls to remain the same, without any motion or alteration? (*T*/54)

If we did not conceive the idea of a vacuum, we would be forced, like 'some metaphysicians', to imagine that the walls of the emptied chamber collapse without moving. But

> I defy these metaphysicians to conceive the matter according to their hypothesis, or imagine the floor and the roof, with all the opposite sides of the chamber, to touch each other, while they continue in rest, and preserve the same position. (*T*/55)

Thirdly, motion would be not only inconceivable, but also altogether impossible, 'without a vacuum, into which one body must move in order to make way for another' (*T*/55). However, Hume declares that he does not want to spend too much time on this third argument, since it 'principally belongs to natural philosophy, which lies without our present sphere' (*T*/55). These classic arguments appear, in various versions, in nearly every discussion of the conceivability of empty space: in Locke's *Essay* and in Bayle's *Dictionary* ('Zeno' and 'Leucippus'), in the *Art de penser*, in Barrow's *Lectures*, and in the works of all sorts of Newtonian writers.

At the beginning of the discussion, Hume follows his usual tactic of referring to the first principle, the priority of impressions over ideas:

> In order to answer these objections, we must take the matter pretty deep, and consider the nature and origin of several ideas, lest we should dispute without understanding perfectly the subject of the controversy. (*T*/55)

What follows uses once more the rhetoric of experimental and natural philosophy, and is variously spiced with half-hearted references to natural philosophical doctrines popular at the time, for instance to the animal spirits of Malebranchean physiology and to the rays of light of geometrical optics.[4] The main point of Hume's argument is that the absence of visual or tactile impressions, being a mere negation, does not and cannot give rise to the idea of a vacuum or extension without matter. ''Tis evident the idea of darkness is no positive idea, but merely the negation of light, or more properly speaking, of colour'd and visible objects' (*T*/55): in utter darkness we would have the same spatial perception as a man born blind, that is, none at all, for 'such-a-one has no idea either of light or darkness' (*T*/56). Similarly, one cannot get any impression of an empty space from the experience of a free motion of our limbs, nor could one get it from being 'softly convey'd along by some invisible power' in a free and invariable motion. The same also applies if the darkness and the motion are 'mix'd with something visible and tangible' (*T*/56): the only difference being that, in this case, we perceive the distant visible and tangible objects, the distance between them being 'known only by the manner, in which the distant objects affect the senses (*T*/59).

In the latter case, however, Hume also finds the key to the illusion of having an idea of empty space. For visible (or tangible) objects that are separated by an interval of darkness (or of free motion of one's hand) appear as a configuration constituted both by the objects and by their distance. In this sense they are related in three ways to objects identically situated, but separated by a visible (or tangible) distance. First, no matter whether or not the interposed distance is perceivable, the

[4] Wright, *Sceptical Realism* (which is a classic of the interpretation now called the 'new Hume'), contains, among other things, a fascinating study of the reception of the Malebranchean animal spirits in Britain.

distant objects 'affect the senses in the same manner, and form the same angle by the rays, which flow from them' (*T*/58; but, in the Appendix, *T*/636, this is taken back). Secondly, the unperceivable distance between two objects may be filled with perceivable objects, or, as Hume puts it, 'an invisible and intangible distance may be converted into a visible and tangible one, without any change on the distant objects' (*T*/59). Thirdly, the two kinds of distance 'have nearly the same effect on every natural phenomenon' and 'all qualities, such as heat, cold, light, attraction, etc., all diminish in proportion to the distance' (*T*/59). When two ideas are closely connected, for instance because they resemble each other, the mind tends to confuse them. The reason may be found in the uncertain motion of the animal spirits; but 'the phenomenon may be real, tho' my explication be chimerical', adds Hume cautiously (*T*/60). The most spectacular case of this confusion depending on close connection is 'That 'tis usual for men to use words for ideas, and to talk instead of thinking in their reasonings' (*T*/61–2).[5] The invisible distance between two luminous objects, which is only an aspect of a perceptual configuration, is closely connected to a distance which is also perceivable in its own right. As a result, we naturally tend to confuse them.

The ground has now been prepared for Hume's answer to the three objections. The first is dissolved: if we commonly take words for ideas, it is obvious that we don't need to have an idea of a vacuum to talk about it. The Cartesian paradox of the emptied chamber whose walls should collapse is also dismissed in a few words:

> When every thing is annihilated in the chamber, and the walls continue immoveable, the chamber must be conceiv'd much in the same manner as at present, when the air that fills it is not an object of the senses. (*T*/62)

As for the natural philosophical argument that without a vacuum motion would be impossible, we know by experience that two bodies separated by an invisible and intangible distance have

> a capacity of receiving body betwixt them, and that there is no obstacle to the conversion of the invisible and intangible distance into one that is visible and tangible. (*T*/63)

For all we know, motion has the same effect on distant bodies as such a

[5] Cf. *Art de Penser*, Part I, chapter 11: 'D'une autre cause qui met de la confusion dans nos pensées et dans notre discours, qui est que nous les attachons à des mots' [in Baynes' transl., p. 76: 'Of another cause which introduces confusion into our thoughts and discourses, that is, that we attach them to words'], and chapter 12.

conversion; there is no need to suppose any empty space. The no-nonsense briskness of this line of thought is breath-taking. The most obvious objection to it is anticipated: should anyone complain that these arguments do not say anything at all about the existence of empty space,

> I answer this objection, by pleading guilty, and by confessing that my intention never was to penetrate into the nature of bodies, or explain the secret causes of their operations. (*T*/64)

The 'science of human nature' is only interested in 'the manner in which objects affect the sense'. The conclusion of the section is the paradox quoted a few pages ago: any discussion about the existence and conceivability of a vacuum is pointless, for positing a vacuum does not offer any escape from absurdity in our description of motion; and whether to use the word 'vacuum' is, on the whole, a matter of taste (*T*/64). In the addition appended in 1740 to Book 3, Hume states his paradoxical position and his scepticism even more openly:

> if it be ask'd, whether or not the invisible and intangible distance be always full of body, or of something that by an improvement of our organs might become visible and tangible, I must acknowledge, that I find no very decisive arguments on either side; tho' I am inclin'd to the contrary opinion, as being more suitable to vulgar and popular notions. (*T*/639)

And the reader is informed that this 'modest scepticism' and the consequent 'fair confession of ignorance in subjects, that exceed all human capacity' are most suitable to '*the Newtonian* philosophy'. (*T*/639)

'These "deep" arguments seem confused', objects John Laird irritably: for instance, after Locke everyone knew that privative causes like darkness can give rise to positive impressions; and some of Hume's remarks are so weak that he takes them back himself having just put them forward.[6] Also, Hume says that the 'idea of darkness' is 'no positive idea', but he never defines what a positive idea is, complains Constance Maund.[7] And, to mention another very obvious problem, the identification between (the idea of) space and (the idea of) extension slips in without any justification. Difficulties of this kind can readily be multiplied. The author of the *Treatise* is famously lacking in good philosophical manners. Equally serious, in my opinion, is the fact that Hume's argument is heavily based on the unclear notion of 'distance',

[6] J. Laird, *Hume's Philosophy of Human Nature*, London 1932, p. 73. The reference to Locke is to *Essay*, II.viii.sections 4ff.

[7] C. Maund, *Hume's Theory of Knowledge*, London 1937, p. 208.

the account of which is, in turn, entirely dependent on the theory that the idea of space derives from the 'manner of appearance' of visual and tactile perceptions. I have suggested in Chapter 2 that the notion of 'manner of appearance' is an excellent way to account for the origin of the idea of space; here, the manner of appearance is alone responsible for our conviction that we can conceive the idea of an empty space. But with the notion of distance the ambiguity latent in the treatment of the 'manner of appearance' becomes manifest – in fact, it becomes so apparent that one may wonder if it is still tolerable: the manner of appearance is at once necessary to account for our conception of the idea of space, and still not sufficient to give rise to a true idea. This looks pretty bad, and, given all the problems left open, the last two pages of the section on the science of human nature sound impossibly smug and arrogant, a triumph of metaphysical irresponsibility.

But, in my opinion, the main problem with this section, what is specially unclear, is to be located at a more basic level. To put it in its crudest form: since it turns out that the whole debate about a vacuum is based on a misunderstanding; since Hume's aim is not that of 'penetrating into the nature of bodies'; and since, in fact, Newtonianism itself, according to Hume, is nothing but a way of describing moving bodies without making any claim to explain the 'secret causes of their operations', why then does Hume engage in this discussion? Given that the conclusion is that the scientist of human nature need not be concerned with such problems as the possibility, existence, and conceivability of a vacuum, why does Hume address them at such length? And, since the idea of a vacuum is inconceivable, what does he think he is talking about?

To show that Hume knows what he is talking about, and to find out what this is, are among the central aims of the present chapter. But let us leave these questions aside for the time being, and approach the problem from a different angle. The only thing that seems clear is that the talk about distance is at the core of Hume's discussion. Distance is the key notion that organises all the imagery of this section – coloured patches on the retina, extensions that are seen and invisible intervals between visible bodies, luminous points against a dark background, darkness and the perceptions of a blind man. All these combine into a strange, unfamiliar way of treating the matter. Here we approach another reason for the difficulty of this text, something that is more subtle than anything else mentioned so far: Hume's pages on the inconceivability of a vacuum give the impression of being remote and alien.

This feeling of alienation may be the reason why Hume scholars, with few exceptions (most of whom have already appeared in my footnotes), have shown a remarkable lack of interest in the conceivability of a vacuum; indeed, in some cases they seem to be at pains to avoid mentioning the subject altogether. In such a lively and crowded field as Hume studies, this is clearly abnormal. My suggestion is that this abnormality has to do with an interpretative frustration which has been discouraging rather than challenging. Let us try to sort out this matter, and see if it is possible to make Hume's discussion appear less odd, and more readable.

AN EVEN MORE DIFFICULT PIECE OF WRITING

As I have observed, Hume's approach to the problem of the conceivability of empty space appears alien. Now, I am going to follow a maxim of practical wisdom according to which, if a text seems impervious to interpretation, it is a good idea to have a look at something similar – close in time, background, language, questions addressed – but even more difficult and alien.[8] Accordingly, in this section I shall consider text which is genuinely alien, that is, two of Law's footnotes to chapter 1 of King's *The Origin of Evil*, 'Containing some Principles necessary to be known in order to the Understanding and Solution of the difficulty about the *Origin of Evil*'.[9] Then, to make a sort of triangulation possible, I shall go briefly into another, less obscure texts; finally I shall return to Hume's *Treatise*.

Law's additions to King's chapter 1 are particularly numerous and long, because of the many points of disagreement between the author and the editor. William King, late Archbishop of Dublin, had been a metaphysical and moral optimist and a negative theologian, with interesting but highly controversial views about knowledge by analogy, in particular about our analogical knowledge of God.[10] Edmund Law, future Master of St Peter's College in Cambridge, then Bishop of Carlisle, was, at the time of his first edition of King's book (he edited five

[8] I owe this principle, however indirectly, to Sue Morgan.

[9] *An Essay on the Origin of Evil.* By Dr. William King, late Lord Archbishop of Dublin. Translated from the Latin, with Large Notes . . . London 1731. The quotations in the text are from this edition.

[10] For King's life see *Dictionary of National Biography*, s.v. Among his writings it is worth remembering the sermon on predestination and foreknowledge, preached on 15 May 1709, which contains a particularly clear and open account of King's doctrine of knowledge by analogy. The essay *De Origine Mali*, his main work, appeared in Dublin and London in 1702 and was widely read both in Britain and on the Continent.

of them in 1731, 1732, 1739, 1758, and 1781), a young Fellow of Christ's College, a follower of Locke's philosophy, and strongly opposed to the version of Newtonianism and the new philosophy put forward by Dr Samuel Clarke and his entourage.[11]

The idea of space appears in King's chapter I in one of the steps towards the demonstration of the necessity of the first cause. The idea of space is particularly interesting to King, who devotes several paragraphs to it, because while motion and matter are clearly not self-sufficient, but dependent on something else as their cause, space 'seems at first sight inseparable from Existence' (p. 30). One of the reasons for this is that we tend to identify ourselves with our bodies, which require space as their 'without' (p. 31); another is that we cannot annihilate in thought parts or portions of space. The situation King uses to illustrate this point is a familiar one:

> We can in Thought remove all Matter out of a Vessel, or Chamber, and the Space interjacent between the Walls remains extended in length, breadth, and depth: But the Space cannot be *removed*, since it is of its own Nature immoveable, nor can it be *annihilated*; for distance would still remain between the bounds which cannot be without Extension, nor Extension without a Subject; but Space, as far as we can conceive it, is the primary Subject of Extension. (pp. 37–40)

The problem of the conceivability of an empty space appears mainly in two of Law's footnotes to King's text. Footnote 5 is attached to the first, rather cursory appearance of the idea of space in King's text. King is setting out his discussion by listing the properties of space as the residue which remains after the annihilation of matter:

> It remains therefore that *Space* (as we conceive it) be something extended, immoveable, capable of *receiving* or *containing* Matter, and penetrable by it. Tho' therefore we have not a *Definition* or *Idea* of *Space*, properly so call'd; yet we can

[11] For Law's life see *Dictionary of National Biography*, s.v. Law's other and more famous works are the *Enquiry into the Ideas of Space, Time, Immensity and Eternity* (1st edition Cambridge 1734), and the *Considerations on the State of the World with regard to the Theory of Religion* (1st edition Cambridge 1745). Both works are independently conceived, but connected to the frequently quoted *The Origin of Evil*. For comments see L. Stephen, *History of the English Thought in the Eighteenth Century*, 2nd edition, London 1902, in which *The Origin of Evil* is also briefly referred to several times (for instance pp. 109, 114ff.), and E. Cassirer, *Die Erkenntnisproblem in der Philosophie und Wissenschaft der neueren Zeit*, Book VI, vol. 2, Berlin 1907, pp. 368ff. on Law's *Enquiry* (a mention of Law's edition of King's *Origin of Evil* is in p. 693, fn. 26). On the general background of Law's work see also J. Hunt, *Religious Thought in England from the Reformation to the End of Last Century. A Contribution to the History of Theology*, London 1870–3; R. T. Holtby, *Daniel Waterland 1683–1740. A Study in Eighteenth-Century Orthodoxy*, Carlisle 1966; and J. Gascoigne, *Cambridge in the Age of Enlightenment: Science, Religion and Politics from the Restoration to the French Revolution*, Cambridge 1989.

hereby sufficiently distinguish it from every other thing, and reason about it as much as we have occasion. (p. 9)

And Law appends a long footnote (five and a half pages in small print) where, for a start, he comments:

Tho' much noise has been made about *Space*, which *Leibnitz* justly calls an Idol of some modern *English* Men; and so great use has been made of it in demonstrating the divine Attributes, in a way which some style *a Priori*; yet, I'm forc'd to confess that I cannot possibly frame any other Notion of it, than either, first, as the mere *negation* or *absence* of *Matter*, or secondly, as the extension of *Body*, consider'd *abstractly*, or separate from any particular Body . . . or thirdly, as a *Subject* or *Substratum* of the same *extension in abstracto*. (p. 9)

The reference to Leibniz's being right in calling space 'an Idol of some modern English men' and to the great use of it 'in demonstrating the Divine Attributes' is transparently a sharp poke at Dr Clarke and his celebrated 1704 Boyle Lectures;[12] these few lines are enough to introduce both Law's style and the metaphysical and theological side he is representing in his work. In particular, attributing infinity to space is – this is the main point Law is making in this footnote – an example of the 'Confusion attending Subjects of this kind', that has been occasioned 'by jumbling Mathematics and Metaphysics together' (p. 15). The potential infinity of mathematics is 'perpetual *Increaseableness* or *Addibility*'; thus it has nothing to do with, and is, in fact, the opposite of, the actual infinity of metaphysics. Law maintains, following Cudworth, that 'Metaphysical Infinite means *perfect*, or *that to which nothing can be added*': it is the sublime perfection of God (pp. 12–13).

Among the three possibilities Law lists – space as privation of matter, as abstract idea of the extension of body, as substratum of the abstract idea of extension – the first is the most interesting for present purposes. If space is nothing but absence or negation of matter, Law writes,

we may indeed have a *positive Idea* of it, as well as of *Silence*, *Darkness*, and many other Privations; as Mr. *Locke* has fully proved that we have, and shewn the Reason of it. B.2 C.8. §.4. But to argue from such an Idea of Space, that Space itself is something external, and has a real existence, seems altogether as good Sense as to say, that because we have a different Idea of *Darkness* from that of *Light*; of *silence* from that of *sound*; of the *absence* of any thing, from that of its *Presence*; therefore Darkness, *&c.* must be something positive and different from Light, *&c.* and have as real an Existence as Light has. (p. 9)

[12] Published in 1705 in London with the title *A Demonstration of the Being and Attributes of God*, and frequently reprinted. Leibniz's remark is in his third letter to Clarke, section 2.

So, we can have as positive an idea of space as of darkness, though both are 'privations'. The reference is to Locke's treatment of privative causes, neatly summarised by Law:

(For every Idea, as it is a perception of the Mind, must necessarily be positive, tho' it arise from what *Locke* calls a privative cause) . . . There are therefore Ideas and *simple* ones too, which have nothing *ad extra* correspondent to them, no proper Ideatum, Archetype, or objective reality, and I don't see why that of space may not be reckon'd one of them. (p. 9)

The idea of space is a positive idea; it is a 'perception of the mind'. After having shown how space cannot have true properties, such as, crucially, that of being extended, Law considers an interesting objection:

Well, but tho' we can't assign any positive or consistent properties to space, yet, say they, we have a Clear Idea of *Distance* itself, and of *different* distances one greater than another, which proves that this is something real. Let us examine our Idea of *Distance* a little farther, and I believe, we shall find it to be only the Conception of a mere *possibility* for a line of such a certain *length*, or a Body of such *dimensions*, to be contain'd between two other Bodies, by mentally applying a *material measure* of so many inches, feet, *&c.* to the supposed *distance* between these two bodies; of which *material measure*, and of its *increasableness*, we have, indeed, a clear Idea, as also of the possibility of its being placed between these two bodies, tho' at present it be not so placed: And this seems to me to be all the conception or Idea, that we can possibly frame about it. (p. 10)

'Distance', according to Law, is another 'merely negative', 'pretended' property, like 'Penetrability, or a capacity to receive Body': in the same way darkness 'has the power or property of *receiving Light*', silence that 'of *admitting Sound*', and absence that 'of being supply'd by *Presence*'. (p. 9)[13]

'They who admit of *Space*, or a *Vacuum* cannot deny but Matter is at least mentally separable from Existence' (pp. 25–6), King writes, thus giving Law the occasion for his footnote 11 (about four pages in small print), specifically devoted to empty space. Law illustrates the discussions about the existence of a vacuum with two long quotations from

[13] See also Law's *Enquiry*, p. 76: 'Mr. *Jackson* concludes his Arguments for the Reality of Space with the old *Cartesian* Plea against a *Vacuum*, *viz.* "If the *Space* or *Distance* between two Sides of a Room be nothing, the Sides would join together, *i.e.* they would not be *distant*, having nothing between them." We deny the Consequence, and assign this plain Reason, *viz.* the Ideas of joining and not joining are really different ideas, and therefore want nothing to hinder them from coinciding or to make them not the same idea. However, the Interposition of another Idea will keep up the Distinction between them without any thing more. I mean, if we place a mere *ideal* or imaginary Object betwixt two others, it will prevent our conceiving the two others to coincide or become one, as much as a real *Object* would; or if we suppose even a bare *Possibility* of placing any thing between them, it will do the same: so far it is from being necessary that we should conceive the two Sides of a Room to touch whenever we suppose no real thing between them'.

Robert Green[14] and from Bayle's 'Zeno'[15], whose arguments 'overthrow the Reality of absolute, simple Space', even though they are 'mere quibbles with regard to the End for which these Authors seem to have urged them'. Discussions of this kind, Law maintains, are similar to those about infinity: they 'may serve as an Instance of the great confusion caus'd by a jumble of Mathematics and Metaphysics together' (p. 26). I shall not here go through either Green's or Bayle's arguments; I shall consider them simply as elaborately sophistical arguments, convenient for Law to wield against his opponents. I shall cite only what Law says before introducing them:

These two Words, *Space* and *Vacuum*, tho' they ought perhaps to have both the same meaning, *i.e.* neither of them to mean any real thing or quality existing in Nature, but only a Negation of Matter and its Qualities, yet as the former is more evidently a positive term, it is apt to convey an Idea of something positive, and thereby lead us to frame some imagination of that something, and so at length draw us into a Notion quite different from that, which the latter Word more naturally offers, and which comes nearer to the truth of the Case; and therefore I think it not quite so convenient to use these two Words promiscuously. It may be doubted whether our substituting the former of these terms for the latter, when the Ideas usually fixt to them have in reality nothing to do with one another, may not have given rise to most of the Disputes against a *Vacuum*, which have been carried on by many able Writers. *Vacuum*, in Natural Philosophy, is (according to the true import of the Word) only *Emptiness*, or absence of Matter, *i.e.* a Term that implies mere Negation; tho', when we come to prove that Matter exists not every where, or that there is really any such emptiness or absence of Matter, we are oblig'd, thro' the Defect of Language, to make use of positive Terms about it, *viz.* that there is a *Vacuum* in this or that Place, or that there is a real Foundation in Nature for supposing it, *&c.* Hence, probably, Metaphysicians, when they come to consider it, being used to contemplate real Essences, are led to understand it, as something positive; which might properly be said to *be here* and *there*, *&c.* Their next Step is, to bring it under the *Imagination*, and so finding the Idea of Space, or Extension, in some measure connected with this Emptiness, they easily substitute one for the

[14] Robert Green, *The Principles of Natural Philosophy. In which is shewn the Insufficiency of the Present Systems, to give us any Just Account of that Science: And the Necessity there is of some New Principles. In Order to furnish us with a True and Real Knowledge of Nature*, Cambridge 1712. On Green, Fellow of Clare Hall, see *Alumni Cantabrigenses*, s.v., and *DNB*, s.v.

[15] Law reports in full a passage from note FDD (vol. 4, pp. 3083–4 in the English edition published in London, 1710), and refers the readers to p. 3077 (containing the end of note E and the beginning of note F), and to note I of the entry 'Spinoza', (*ibid.*, p. 2790). Kemp Smith, *Philosophy*, Appendix C to chapter xiv, pp. 325–38, esp. pp. 331–8, reports many relevant passages, including part of those cited by Law (p. 335) (but in the English translation published in London, 1738, with note I being the same as note FDD in the 1710 edition, etc.). On Hume's use of Bayle see also Laird, *Philosophy*, pp. 78–9.

other, and often change the negative Idea into a positive one, and define *Vacuum* to be *Extension void of Solidity*, or *Space without Body* (Locke, B.2. C. 13. §. 22), Whereas the Ideas of Vacuity and Extension have no real Connection with each other, as was said before, tho' they be very apt to go together. Well then, these two distinct Ideas being both included under the Word *Vacuum*, it becomes equivocal, and consequently that may be affirm'd or deny'd of it according to the one Idea, which cannot according to the other, and here is room for endless Juggle. *v.g.* It may be said that there is a real foundation in Nature for supposing a *Vacuum* in the negative sense of the Word, *i.e.* as signifying mere emptiness; but the same thing may be denied of it in the positive, *i.e.* standing for pure extension, which is an *abstract Idea*, form'd by the mind itself, and, as such, has no foundation any where else. Again, Philosophers, who take a Vacuum for Space or Extension in the abstract, stifly deny that there is a Vacuum in Nature, which is true indeed of absolute Space, which exists only in the Mind, but is not so of Vacuity or absence of Matter, which has as real a foundation in nature as Matter itself has: except we'll argue that it cannot be said to *be* or to have *existence* predicated of it, because 'tis only a negation; which is playing upon, and puzzling one another with Words. (pp. 25–6)

Law's emphasis is all on words and on their effect on the mind. Words describe the perceptions of the mind, and the ease with which the mind moves from one idea to another gives rise to illusions. The idea of a vacuum is an illusion created by the effect of words on our minds.

Law's interminable, dense, and now nearly incomprehensible footnotes are typical of a whole tradition of thinking and writing. From this point of view, some external aspects of *The Origin of Evil* are also of crucial interest: for instance, this interpretation of the way of ideas appears not in an independent text, but as footnotes and additions to somebody else's work. Also, it is remarkable that in these footnotes one can hardly find their author – they are a true forest of quotations, from Gassendi, Raphson, and Dr Clarke on the infinity of space, and from Cudworth and Waterland against it, from Green and Bayle against absolute space, and so on. Even when he is not quoting, this author's writing often has the distinct flavour of someone else's: for instance, the passage quoted above from the footnote on vacuum sounds very much like Waterland.[16] In other words, *The Origin of Evil* is a specimen of a very

[16] Law's eclecticism is deliberate. As he says in the Preface: 'Some perhaps may think the frequent and long Quotations very tedious, and introduced only to stuff up. I can only answer, that I intended by the Notes, and References together, to point out a sort of *Compendium of Metaphysics*, or Speculative Divinity; but directing the Reader to a set of true Notions on the various Subjects which our Author touch'd upon; and which could not be found in any one particular Book, nor collected from several, without much Trouble and Confusion, and unnecessary Reading. I chose rather to quote the very Words of the Authors, than either use worse of my

well-defined genre. Concerning the discussion of the conceivability of vacuum, I find this document interesting not because its difficulty is the sign of an original approach, but, on the contrary, precisely because all the *loci communes* of the discussion on the conceivability of empty space appear in it.

And yet *The Origin of Evil* was a highly controversial book. In order to appreciate fully both the idiosyncrasies and the standard features, I propose a brief tour of another text, this time a very orthodox one (in genre as well as in content) – Chambers' *Cyclopaedia*. It will appear that again God has an important part to play.

It is possible to step into Chambers' labyrinth from the entry 'Vacuum', where one finds a different terminology, inherited from 'the antients', who 'distinguished two kinds; a *vacuum coacervatum* and a *vacuum interspersum* or *disseminatum*'.[17] I am interested in the first, that is defined as 'a place destitute of matter: such, *e.g.* as there would be, should God annihilate all the air, and other bodies within the walls of this chamber':

> According, then, to [the modern corpuscularians] there is no *vacuum coacervatum* without the bounds of the sensible world: nor would there be any vacuum, provided God should annihilate divers contiguous bodies, than what amounts to a mere privation, or nothing: the dimensions of such a space, which the antients held to be real, being by these held to be mere negations; that is, in such a place, there is so much length, breadth and depth wanting, as a body must have to fill it. – To suppose, that when all the matter in a chamber is annihilated, there should yet be real dimensions, is to suppose corporeal dimensions without body; which is absurd. The Cartesians, however, deny any *vacuum coacervatum* at all; and assert, that if God should immediately annihilate all the matter, *v.g.* in this chamber, and prevent the ingress of any other matter, the consequence would be, that the walls would become contiguous, and include no space at all. – They add, that if there be no matter in a chamber, the walls

own, or pretend to discover what had been often discovered before; or repeat the same things over and over again, which is endless. I hope the Reader will find that a citation of two or more Authors on the same Point is not always tautology: and I believe it will appear, that in the multitude of *References* more than one is scarce ever made to the same Place, except upon a very different Occasion, or in some different Light. A Writer seems to me to do more good to the public by shewing the use of some of those many Volumes which we have already, than by offering new ones; tho' this be of much less Advantage to his private Character. I determin'd therefore not to say any thing myself where I could bring another conveniently to say it for me; and transcribed only so much from others as I judg'd absolutely necessary to give the Reader a short view of the Subject we were upon, and by that Sketch to induce those who have leisure, opportunity and inclination to go farther, and consult the Originals; and to afford some present Satisfaction to those who have not' (p. x).

[17] Cf. Law, *Origin of Evil*, fn. 12, p. 29: 'from all the proofs of a *Vacuum*, both *external* and *interspersed*, it appears that all Matter is every way limited or defective'.

can be conceived no otherwise than as contiguous, those things being said to be contiguous, between which there is not any thing intermediate: but, if there be no body between, there is no extension between; extension and body being the same thing: and if there be no extension between, then the walls are contiguous; and where is the *vacuum*? But the reasoning is built on a mistake, *viz.* that body and extension are the same thing. See EXTENSION, and SPACE.[18]

Let us consider 'Space':

Space, considered barely in length between any two bodies, is the same idea which we have of distance. See DISTANCE . . . When considered between the extremities of matter, which fills the capacity of *space* with something solid, tangible and moveable, it is called *extension*. See EXTENSION . . .

Distance

properly denotes the shortest line between two points, objects, *&c.* The word is also used figuratively for an interval, not only in respect of place, but also of time, or quality . . .

Under 'Extension' we find:

If a man considers the distance between two bodies, abstractedly, and without any regard to bodies which may fill that interval, it is called *space*: and when he considers the distance between the extremes of a solid body, it is called *extension*. See SPACE . . .

Going back to 'Space':

. . . extension is an idea belonging to body only; but *space*, it is plain, may be considered without it. See BODY and VACUUM. *Space*, therefore, in the general signification, is the same thing with distance considered every way, whether there be any solid matter in it, or not. Each different distance is a different modification of *space*; and each idea of any different *space* is a simple mode of this idea.

To 'Vacuum' again: if there is no vacuum,

all motion must be in circles geometrically true; and the revolving bodies must be either spheres, spheroids, cylinders, or proportions of them, exactly geometrical; otherwise, their revolutions in a plenum would be impossible: but such motions, or such figured bodies, we do not know in nature. Therefore there is a *vacuum*.

And Chambers explains at length all the pros and cons with regard to the motion of planets, the weight of bodies, the descent of bodies, the vibrations of a pendulum, the actual division of matter in parts, the

[18] All quotations are from the edition published in London in 1738.

figure of bodies, and the infinity of the material world if one denies vacuum. Under 'Plenum' one finds:

PLENUM, in physics, is a term used to signify that state of things, wherein every part of space, of extension, is supposed to be full of matter. See MATTER and SPACE. In opposition to a *vacuum*, which is a space supposed devoid of all matter. See VACUUM. The Cartesians adhere firmly to the doctrine of absolute *plenum*. This they do on this principle, that the essence of matter consists in extension; from whence, indeed, the consequence is very easy, that where-ever there is space or extension, there is matter. See EXTENSION. But this principle has been shown to be false; and therefore the consequence drawn from it falls to the ground. See MATTER. But that there is a real vacuum in the nature of things, is likewise demonstrated by arguments *a posteriori*, under the article VACUUM.

In conclusion, back to 'Vacuum' once more:

Since then the essence of matter does not consist in extension, but in solidity, or impenetrability, the universe may be said to consist of solid bodies moving in a *vacuum*: nor need we at all fear, lest the phenomena of nature, most of which are plausibly accounted for from a plenitude, should become inexplicable when the plenum is set aside. The principal ones, such as the tides; the suspension of the mercury in the barometer; the motion of heavenly bodies, of light, *&c.* are more easily and satisfactorily accounted for from other principles.

And thus the circle is closed.

It is now time to assess the results of this joint reading of Law and Chambers. For instance, the focus on the key-notion that I have chosen – distance – is common to both texts. In fact, one may take it to be a commonplace, since it appears to be directly derived from one of the most famous Cartesian tenets which is at this point useful to cite in full:

Lapsique sumus ferè omnes à primâ aetate in hunc errorem, propterea quòd, non advertentes ullam esse inter vas & corpus in eo contentum necessariam conjunctionem, non putavimus quicquam obstare, quonimus saltem Deus efficiat, ut corpus, quod vas aliquod replet, inde auferatur, & nullum aliud in ejus locum succedat. Jam autem, ut errorem illum emendemus, considerare oportet nullam quidem esse connexionem inter vas & hos vel illud corpus particulare quod in eo continetur, sed esse maximam, ac omnino necessariam, inter vasis figuram concavam & extensionem in genere sumptam, quae in eâ cavitate debet contineri. Adèo ut non magis repugnet nos concipere montem sine valle, quàm intelligere istam cavitatem absque extensione in eâ contentâ, vel hanc extensionem absque substantiâ quae sit extensa: quia, ut saepe dictum est, nihili nulla potest esse extensio. Ac proinde, si queratur quid fiet, si Deus auferat omne corpus quod in aliquo vase continetur, & nullum aliud in allati locum venire permittat: respondendum est, vasis latera sibi invicem hoc ipso fore contigua. Cum enim inter duo corpora nihil interjacet, necesse est ut se mutuò tangant; et manifestè repugnat ut distent, sive ut inter ipsa sit distantia, & tamen ut ista distantia sit nihil: quia omnis distantia est modus extensionis, & ideò sine substantiâ extensâ esse non potest.[19]

The definition on the basis of distance is certainly related to other ways of defining space which are also common: for instance, Barrow describes space as '*nothing else but the mere Power, Capacity, Ponibility*, or (begging pardon for the Expressions) *Interponibility of Magnitude*'.[20] Other examples of this sort may be taken from Bayle. The link between the treatment of empty space and that of privative notions is also made in both Law's and Chambers' texts. The first, however, presents it in more explicitly Lockean terms, with the appearance of darkness and the other privative causes. Also, while Law takes a typical and quite sophisticated way-of-ideas approach to the problem of the conceivability of empty space, Chambers seems to be quite happy – as usual – with a general trust in the success of natural philosophy.

The debate to which Law's footnotes and Chambers' entries both belong, however differently, is part of a tradition of discussion that was once most lively and popular and is now lost, and the measure of its being lost is precisely the feeling of alienation that one experiences when faced with many of its documents. Some of them – Locke's *Essay*, Hume's *Enquiries*, perhaps even some parts of his *Treatise* – may not

[19] *Principia Philosophiae*, II, 18 (in the English translation of Haldane and Ross: '*How the prejudice concerning the absolute vacuum is to be corrected.* We have almost all lapsed into this error from the beginning of our lives, for, seeing that there is no necessary connection between the vessel and the body it contains, we thought that God at least could remove all the body contained in the vessel without its being necessary that any other body should take its place. But in order that we may be able to correct this error, it is necessary to remark that while there is no connection between the vessel and that particular body which it contains, there is an absolutely necessary one between the concave figure of the vessel and the extension considered generally which must be comprised in this cavity; so that there is not more contradiction in conceiving a mountain without a valley, than such a cavity without the extension which it contains, or this extension without the substance which is extended, because nothing, as has already been frequently remarked, cannot have extension. And therefore, if it is asked what would happen if God removed all the body contained in a vessel without permitting its place being occupied by another body, we shall answer that the sides of the vessel will therby come into immediate contiguity with one another. For two bodies must touch when there is nothing between them, because it is manifestly contradictory for these two bodies to be apart from one another, or that there should be a distance between them, and yet that this distance should be nothing; for distance is a more of extension, and without extended substance it cannot therefore exist'; (*The Philosophical Works of Descartes* rendered into English by E. S. Haldane and G. R. T. Ross, vol. I, Cambridge 1911, p. 263).

[20] Isaac Barrow, *The Usefulness of Mathematical Learning Explained and Demonstrated; Being Mathematical Lectures Read in the Public Schools at the University of Cambridge* by Isaac Barrow, D.D. Professor of Mathematics, and Master of the Trinity College, etc. Translated by the Revd Mr John Kirkby, of Egrement in Cumberland, London 1734 (facsimile edition London 1970), p. 175.

initially feel either 'lost' or particularly alien. This is because, however absent-mindedly and indirectly, we are heirs to an uninterrupted tradition of familiarity with them: it is partly a sort of optical illusion (for they do feel alien when one tries to approach them in their own terms, as Hume's pages on the idea of space show with special clarity) and partly a sign of their survival in our tradition. They have been carried over, in a mutilated and more or less creatively misinterpreted state, into our culture. But the feeling of alienation inspired by such writings as Chambers' *Cyclopaedia* is very strong. And, in the case of Law's edition(s) of King's essay, it has been and is so overwhelming that the fact of its having been a very important book – for some decades among the crucial readings in Cambridge and at the centre of authoritative comments and reviews, learned disputations and polemics, some gossip, and a bit of scandal – has not been enough to persuade historians of philosophy to discuss it.[21] In any case, Hume knew of it: in fact, he may have read it, since it is quoted in his early notes and in the *Dialogues*.[22]

Bits and pieces of Hume's approach to the problem of the conception of a vacuum seem to appear both in Chambers and in Law – certainly, the terms of reference now seem less inscrutable. The language used in the *Treatise* is all there. Moreover, some very Humean features of Hume's discussion appear already in these texts, especially in Law's notes: a habit leading the mind in a certain direction, a certain resemblance between two terms making it easy to mistake one for the other, and the sharp distinction between natural philosophy and metaphysics. It is important, however, to notice that there is a very clear-cut difference which marks off the writing of the author of the *Treatise*: the absence of God. This is not particularly surprising – Hume's meta-

21 There are some obvious exceptions: for instance, Stephen and Cassirer (as above in fn. 11). In connection with Hume's philosophy, *The Origin of Evil* is briefly referred to in J. Laird, *Philosophy*, pp. 98, 302ff.; R. H. Hurlbutt, *Hume, Newton and the Design Argument*, Lincoln 1965; and D. Cupitt, *Christ and the Hiddenness of God*, London 1971, pp. 69ff.

22 The quotation in the *Dialogues* is in Part 10, fn. 1, p. 194. For Hume's early notes, see Mossner, 'Hume's early memoranda, 1729–1740: the complete text', *Journal of the History of Ideas* 9, 1948, pp. 492–518. Mossner suggests that 'some of the intervening Bayle references are actually 'lifted' from the copious notes added to [the] translation by Edmund Law'. His conclusion is that 'It is not unlikely, to be sure, that Law directed Hume's attention to some important passages in Bayle. By and large, however, a citation by Hume carries its own guarantee that he was actually reading the work named' (p. 496). On the other hand, J. P. Pittion, in his excellent paper 'Hume's reading of Bayle: an inquiry into the source and role of the Memoranda', *Journal of the History of Philosophy* 15, 1977, pp. 373–86, considers the opposite possibility and suggests that Hume may have not read King's book at all, but have known about it either through Bayle, or through some journal review. Mossner's authority is beyond discussion; but I also find Pittion's suggestion sensible. Of course, the resolution of this question does not affect the point I intend to make.

physical interest in God is famously of a different sort – and yet, it deserves further comment.

God invariably appears in these discussions about the existence or the conceivability of empty space. In Law's footnotes, it is God's sublimity that gives a sense to the whole structure. In Chambers, the appearances of God are frequent and relatively sober, as in Descartes' passage at the beginning of the whole story (where, as I have said, he wonders what would happen, 'si Deus auferat omne corpus quod in aliquo vase continetur, & nullum aliud in allati locum venire permittat'). A particularly dramatic appearance of God in a discussion of empty space is to be found in Barrow:

> If all the Matter be excluded by the divine Power from between these Walls, there will actually be no Body between them, i.e. *there is an intermediate Space.*[23]

In Locke, the appearances of God sometimes start in a similar way, but then are transformed:

> those who assert the impossibility of *Space* existing without *Matter*, must not only make Body infinite, but must also deny a power in God to annihilate any part of Matter. No one, I suppose, will deny, that God can put an end to all motion that is in Matter, and fix all the Bodies of the Universe in a perfect quiet and rest, and continue them so as long as he pleases. Whoever then will allow, that God can, during such a general rest, annihilate either this Book, or the Body of him that reads it, must necessarily admit the possibility of a *Vacuum.*[24]

In other passages, the experimental imagination takes over completely and the result is quite amusing. For instance:

> If *Body* be not supposed infinite, which, I think, no one will affirm, I would ask, Whether, if God placed a Man at the extremity of the corporeal Beings, he could not stretch his Hand beyond his Body? If he could, then he would put his Arm, where there was before *Space* without *Body*; and if there he spread his Fingers, there would still be *Space* between them without *Body*: If he could not stretch out his Hand, it must be because of some external hindrance; (for we suppose him alive, with such a power of moving the parts of his Body, that he hath now, which is not in itself impossible, if God so pleased to have it; or at least it is not impossible for God so to move him).[25]

Locke's images of God are very conversible, if somewhat pedantic; this

[23] I. Barrow, *Mathematical Lectures*, p. 176.

[24] *Essay*, II.xiii. section 21 bis. Immobility and annihilation (but again without God) appear in Hume's version of the Cartesian paradox in *T*/54 (and annihilation also appears, coupled with creation and once more without any mention of God, in the solution of the third objection in *T*/63).

[25] *Ibid.*, II.xiii. section 21.

God is a bit whimsical in his experiments and may well incinerate Locke's own work or its reader – even if one may feel that it is not with the man guinea-pig, but rather with the experimenting God, that the reader is invited to identify. In Hume's pages on the conceivability of empty space, God hardly appears at all. For instance, the presentation of the Cartesian paradox is as follows:

> Every idea is possible, which is a necessary and infallible consequence of such as are possible. Now tho' we allow the world to be at present a plenum, we may easily conceive it to be depriv'd of motion; and this idea will certainly be allow'd possible. It must also be allow'd possible, to conceive the annihilation of any part of matter by the omnipotence of the deity, while the other parts remain at rest. For as every idea, that is distinguishable, is separable by the imagination; and as every idea, that is separable by the imagination, may be conceiv'd to be separately existent; 'tis evident, that the existence of one particle of matter, no more implies the existence of another, than a square figure in one body implies a square figure in every one. This being granted, I now demand what results from the concurrence of these two possible ideas of rest and annihilation, and what must we conceive to follow upon the annihilation of all the air and subtle matter in the chamber, supposing the walls to remain the same, without any motion or alteration? (*T*/54)

The emphasis is all on conception and conceivability, and the 'omnipotence of the deity' so minimally invoked does not sound awe-inspired – rather, it is treated as a familiar fiction of the mind. One may find, as it were, traces of God (or, at least, places vacated by God) in Hume's pages. For instance, when he shows that the absence of perceptions does not give rise to any idea:

> Suppose a man to be supported in the air, and to be softly convey'd along by some invisible power; 'tis evident he is sensible of nothing, and never receives the idea of extension, nor indeed any idea, from this invariable motion. Even supposing he moves his limbs to and fro, this cannot convey him that idea. (*T*/56)

The focus of the image is not God, for God has turned into 'some invisible power', and one immediately identifies with the little man kicking around and failing to perceive extension. Hume's position has one major attraction: in place of the metaphysically sublime one finds here the less alarming metaphysically ridiculous. But the price paid for this is high: Hume's approach to the question of empty space belongs to a well-established contemporary tradition, but, once the standard reference to God is eliminated, it is no longer recognisable in the terms of that tradition, hence no longer seems to make sense. The point is that, within the frame of reference that I have illustrated above, the appeal to God is not only conventional or pious, it is an essential part of the whole conception. Should one conclude from this that not only is Hume not very original on this subject, but that he also makes a mess of the approach he is adopting; and that his text is not only unclear, but also positively confused? I do not think so. It is now time to go back to Hume's text and start seriously trying to make sense of it. I shall begin by examining 'distance' and identifying what count as simple spatial impressions.

SIMPLICITY AND DISTANCE

Hume's talk about 'distance' no longer feels totally unfamiliar; it no longer needs excusing. But, if anything, this newly acquired familiarity makes the interpretative predicament seem yet more hopeless. In particular, the elimination of God, the most remarkable difference between Hume's text and the other ones I have considered, has apparently catastrophic consequences for the balance of his discussion. But there is something worse, namely the impression that the whole section may be a futile game: for it appears quite possible that Hume wrote these impenetrable pages unreflectively, without realising he had nothing to say on the issue, simply because there were available standard ways of discussing these matters.[26] So, perhaps he (literally) did not know what he was talking about, but was on automatic pilot – did he write simply as 'un admirable mosaïste et un habile metteur en scène'?[27] I shall suggest at the end that this is not the case; but first let us consider more closely what Hume did write. So, in this section I shall examine Hume's talk about distance in the light of his notion of simple idea. My aim is to show that the invisible and intangible distance which appears in Hume's discussion of the conceivability of empty space is a radicalised version of the 'manner of appearance' of the coloured points composing the table and – space being also the situation we, our bodies, are in – of the objects surrounding our body:

[26] Laird, *Philosophy*, pp. 64–6, while acknowledging Hume's genuine interest in the problems concerning space and time, points out that 'they permeated the thought of Hume's time', so that 'Hume's excursions . . . into these subjects' were 'scarcely avoidable'. And, after a learned and stimulating discussion of Hume's position on infinite divisibility, geometry, and empty space, he proceeds (pp. 78–9) to list Hume's debts to Bayle and to Berkeley, to show that 'Hume's general arguments concerning space and time . . . were less original than is sometimes supposed.'

[27] The expression is André Leroy's (who, of course, rejects this image of Hume): see his introduction to his translation into French of the *Treatise*, *Traité de la nature humaine*, Paris 1946, p. 7.

if it be ask'd, whether the invisible and intangible distance, interpos'd betwixt two objects, be something or nothing: 'Tis easy to answer, that it is *something*, viz. a property of the objects, which affect the *senses* after such a particular manner. (*T*/638)

The unperceivable distance is the manner of appearance projected away from, and made independent of, the objects that appear, and also, as it were, claiming its independence from the perceiver. In this sense, the unperceivable distance is radical in another way as well – it has a more spontaneous and somehow stronger claim (or pretence) to reality:

if it be ask'd, whether two objects, having such a distance betwixt them, touch or not: It may be answer'd, that this depends upon the definition of the word, *touch*. If objects be said to touch, when there is nothing *sensible* interpos'd betwixt them, these objects touch: If objects be said to touch, when their *images* strike contiguous parts of the eye, and when the hand *feels* both objects successively, without any interpos'd motion, these objects do not touch. (*T*/638–9)

This sounds like Hume at his most Pyrrhonian. And yet, Cassirer gives a fascinating account of Hume's position on the idea of a vacuum:

Totally empty space is in and for itself admittedly not a possible object of representation; if, however, we are given two concrete, visual or tangible spatial elements, with respect to which the consciousness possesses a definite sensible substrate, we represent together with them their relative position and their *distance*. For in order to do this we do not need to visualise and fill with actual sensations the whole extent between them; rather, the thought suffices that where we now observe a pure nothing, the *absence* of any perceptual content, a new sensation could appear and take its place. However, it is evident that these observations presuppose the intuition of space rather than clarify it.[28]

Vacuum is absence and cannot be represented. But perceiving two-bodies-separated-by-an-invisible-and-intangible-distance is perceiving a configuration, a whole including the two bodies' mutual position and distance as well as the bodies themselves; the invisible and intangible distance is 'a property of the objects, which affect the *senses* after such a particular manner'. The notion of 'manner of appearance' as

[28] Cassirer, *Das Erkenntnisproblem*, Book 5, pp. 255–6, my transl. ['Der völlig leere Raum ist freilich an und für sich kein möglicher Gegenstand der Vorstellung, wohl aber stellen wir, wenn uns zwei konkrete, sicht- oder tastbare Raumelemente gegeben sind, an denen das Bewusstsein ein festes sinnliches Substrat besitzt, zugleich mit ihnen auch ihr Lageverhältniss und ihre *Entfernung* vor. Denn um dies zu tun, brauchen wir uns nicht die gesamte Zwischenstrecke, die sie trennt, in all ihren Teilen einzeln zu vergengenwärtigen und sie mit wirklichen Empfindungen auszufüllen, sondern es genügt der Gedanke, dass dort, wo wir jetzt ein blosses Nichts, wo wir die *Abgewesenheit* jeglichen Wahrnehmungsinhalt bemerken, eine neue Empfindung auftreten und ihren Platz finden *könnte*. Dass indessen diese Bemerkungen die Anschauung des Raumes voraussetzen, statt sie zu erklären, ist ersichtlich'].

the result of an operation of the mind, that is, as a contribution of the mind to the formation of the idea of space, is familiar already from an earlier chapter. Here Hume states that the idea of space derives from the whole configuration, with the invisible and intangible distance being part of its 'manner of appearance'. Similarly, in the few lines devoted to the origin of the idea of empty time we find:

we may observe, that there is a continual succession of perceptions in our mind; so that the idea of time being for ever present with us; when we consider a stedfast object at five-a-clock, and regard the same at six; we are apt to apply to it that idea in the same manner as if every moment were distinguish'd by a different position, or an alteration of the object. The first and second appearances of the object, being compar'd with the succession of our perceptions, seem equally remov'd as if the object had really chang'd. (*T*/65)

Here there is no opposition between duration and the identity of the unchanging object, since duration only appears when the object is set against the background of our ever-changing consciousness – this is, in fact, the origin of the idea of identity:

'tis only by a fiction of the imagination, by which the unchangeable object is suppos'd to participate of the changes of the co-existent objects, and in particular of that of our perceptions. This fiction of the imagination almost universally takes place; and 'tis by means of it, that a single object, plac'd before us, and survey'd for any time without our discovering in it any interruption or variation, is able to give us a notion of identity. (*T*/200–1)

So, time is projected onto something which does not in itself even afford the occasion for a temporal arrangement of perceptions – the essence of time appears to be consciousness itself.

Contrary to what one may expect, the situation is even more complicated with regard to the idea of space. Let us consider again an already familiar passage:

When I hold up my hand before me, and spread my fingers, they are separated as perfectly by the blue colour of the firmament, as they cou'd be by any visible object, which I cou'd place betwixt them. (*T*/56)[29]

Both the fingers and the space between them look as if painted on a surface. So they appear on the retina: 'distance' is not to be found in the crude sense-impression, but only comes out through an act of judgement. Of course, this is a primordial impression of which normally we are not aware: we immediately and unreflectively decode the painted

29 Note the echo from the passage quoted above from Locke's *Essay*, II.xiii. section 21, where the man placed by God 'at the extremity of corporeal Beings' spreads his fingers in the space without body.

surface of the retina, and 'see' the coloured shapes of bodies as nothing but familiar-looking things which surround us all the time without giving rise to philosophical curiosity. It is the science of human nature which discovers the original impression, a unitary one where the fingers and the blue sky are not differentiated. Pure sense-perception is only grasped by the practised (or perhaps just alerted) eye of the 'scientist of human nature', who must, then, reconstruct experience.[30] Compared with experience as we normally know it, the total impression acquired in an immediate act of perception appears astonishingly simple. This point needs examining in more detail.

Simple impressions play a very important part in Book I of the *Treatise*; and yet (typically) it is far from easy, on the basis of the text, to determine what exactly they are and where one is to look for them. I suppose one may be tempted to choose the coloured points, the elementary units of sight, as typical 'simple impressions', and even to go back to the passage on the grain of sand for a description of a simple impression: 'the idea of a grain of sand is not distinguishable, nor separable, into twenty, much less into a thousand, ten thousand, or an infinite number of different ideas' (*T*/27). Extension consists of 'simple indivisible ideas', Hume writes; a coloured point 'according to the supposition, is perfectly simple and indivisible' (*T*/38).[31] This interpretation is literally acceptable, but it is not unproblematic: the units of sight are very ambiguous and, however certainly 'simple', they are not entirely satisfactory as exemplars of perceptual simplicity. A point is called 'simple' because it is indivisible and elementary; but it is not at all simple if one considers the way it comes to be perceived: in fact, as I have shown, the impressions of grains of sand and coloured points are very complicated, very sophisticated items.[32] I suggest that simple

[30] See above, chapter 2, 'A philosopher's eye'. (Another eye which can grasp purely visual data is, of course, that of the newly sighted of Molyneaux's problem.)

[31] See S. Tweyman, 'Hume on separating the inseparable', in W. Todd (ed.), *Hume and the Enlightenment*, Edinburgh 1974, pp, 30–42, esp. p. 33, and in 'Some reflexions on Hume on existence', *Hume Studies* 18, 1992, pp. 137–49, esp. p. 139. But see also D. Livingston, *Hume's Philosophy of Common Life*, Chicago 1984, esp. pp. 52–3, to which the present discussion is much indebted. See also L. Falkenstein, 'Hume's account of space and time as manners or dispositions', *Archiv für Geschichte der Philosophie* 79, 1997, pp. 179–201. On simple ideas and distinctions of reason see R. J. Butler's important '*Distinctiones rationis*, or the Cheshire Cat which left its smile behind it', *Proceedings of the Aristotelian Society*, N.S. 76, 1976, pp. 165–76, and 'Hume's impressions', in G. Vesey (ed.), *Impressions of Empiricism*, London 1976, pp. 122–36. An interesting recent discussion is in D. Garrett, *Cognition and Commitment in Hume's Philosophy*, Oxford 1997, chapter 3, esp. pp. 60–4 (where the simplicity of a visual unit and the simplicity of the entire impression of a globe of white marble are connected by reference to the type/token distinction).

[32] See above, chapter 2.

impressions are instantaneous perceptions or, as it were, emotional rather than visual units. So the perceptual wholes of a sphere of white marble, of two luminous points on a dark field, or of our fingers spread against the sky, may be better candidates for simple impressions than the grain of sand whose image is identical with that of its thousandth and ten thousandth part.

In other words, a simple impression is defined through its immediacy. Let us consider a famous case, which takes us to the core of the question:

> 'Tis certain that the mind wou'd never have dream'd of distinguishing a figure from the body figur'd, as being in reality neither distinguishable, nor different, nor separable; did it not observe, that even in this simplicity there might be contain'd many different resemblances and relations. (*T*/25)

Hume is introducing the discussion of the distinction of reason. Here, a simple impression makes its appearance in his page explicitly enough:

> Thus when a globe of white marble is presented, we receive only the impression of a white colour dispos'd in a certain form, nor are we able to separate and distinguish the colour from the form. (*T*/25)

The primordial simple experience Hume is referring to is that of the contact with an object that happens before the distinctions of judgement have articulated the world of experience. Once again the language is allusive – the colour is 'dispos'd in a certain form'. Catching a resemblance is, in the words of a very insightful Hume reader, to note the 'unwitting transition of the mind' from an object to another and, in the case of distinctions of reason, to compare not perceptions, but comparisons.[33] What follows makes it clear that the simplicity of this simple impression lies in its coming before the perceiver has learned how to complicate it:

> But observing afterwards a globe of black marble and a cube of white, and comparing them with our former object, we find two separate resemblances, in what formerly seem'd, and really is, perfectly inseparable. After a little more practice of this kind, we begin to distinguish the figure from the colour by a distinction of reason. (*T*/25)

This is, I submit, a phenomenology of the initial stages of perceptual acquaintance with the world. It is the history of consciousness learning how to judge – it is a history because consciousness exists in time and develops, so that, once the ability to judge, to make comparisons and to

[33] Butler, 'Cheshire Cat', pp. 169–70, and 'Hume's impressions', p. 128.

find resemblances has been acquired, there is no way back. As Hume puts it: 'By this means we accompany all our ideas with a kind of reflection, of which custom renders us, in a great measure, insensible' (*T*/25). The 'scientist of human nature' can recover the lost stage of the undifferentiated simple ideas with his well-trained introspection, and hint at it with his metaphorical and allusive language. This is because the 'scientist of human nature' is not only in the habit of judging, that is of observing resemblances as we all do; he has also discovered that this is what he is doing all the time. By this means, he discovers that the notion of 'distance' arises as a result of our judgement's interpretation of visual data. My hand, fingers spread against the background of the blue sky, is apprehended by the eye unassisted by judgement as a whole, unitary impression. As such, it is to be considered as a simple impression; however visually extended and therefore composite, it is an emotional unit.

'Tis evident, that even different simple ideas may have a similarity or resemblance to each other . . . *Blue* and *green* are different simple ideas, but are more resembling than *blue* and *scarlet*' (*T*/637), Hume says about less controversial simple ideas. The resemblance that matters in the case of the abstract idea of space is of the same sort: it is the unwitting transition of the mind from the coloured points 'dispos'd in a certain manner' composing the first extended object one experiences, to the violet, green, red, white, black objects in which we find 'a resemblance in the disposition of the colour'd points, of which they are compos'd'. In the end,

> we omit the peculiarities of colour, as far as possible, and found an abstract idea merely on that disposition of points, or manner of appearance, in which they agree. (*T*/34)

All this seems to refer us back for further light to Hume's treatment of abstract ideas. But it is now time to return to the issue of the conceivability of empty space.

IDEAS AND THE POWER OF WORDS

'Now 'tis certain we have an idea of extension; for otherwise why do we talk and reason concerning it?' (*T*/32), so Hume writes in an earlier section of the *Treatise*. About thirty pages later, one finds that:

> The frequent disputes concerning a vacuum, or extension without matter, prove not the reality of the idea, upon which the dispute turns; there being

nothing more common, than to see men deceive themselves in this particular; especially when by means of any close relation, there is another idea presented, which may be the occasion of their mistake. (*T*/62)

One talks and reasons about extension, therefore extension exists. One talks and reasons about a vacuum; and yet a vacuum is inconceivable, we are just playing with words. One is tempted to say that this lack of coherence is another result of the bad manners of the author of the *Treatise*. That we have an idea of extension is so positively certain, he thinks, that one need not bother to find a decent argument to show it – or perhaps in his attack on the conceivability of a vacuum he is completely carried away by his own polemical intentions. If we stick to the text, however, we may find a more interesting way of putting the matter. A vacuum is inconceivable, in the sense that our idea of a vacuum is a fiction. Still, the whole discussion presupposes that we grant that this fiction is at least good enough to allow conversation – we can keep talking, and we can reasonably expect to be understood: 'vacuum' and 'empty space' are perfectly good words, and in the *Treatise* itself twelve pages are devoted to discussing the inconceivable idea of empty space or a vacuum. One may wonder why Hume talks about them at all, and how he thinks he can do so if they are inconceivable – but his discussion is there, so I suggest that it may be a good idea to consider how, in the *Treatise*, talking is supposed to work and what words are. A good starting-point is to be found in Part 1, section 7, 'Of abstract ideas'.

The content of this section is well known. The crucial question with which Hume begins is '*whether they* [abstract ideas] *be general or particular in the mind's conception of them*' (*T*/17). The answer is a famous creative appropriation of Berkeley's theory of abstract ideas:

all general ideas are nothing but particular ones, annexed to a certain term, which gives them a more extensive signification, and makes them recall upon occasion other individuals, which are similar to them. (*T*/17)[34]

In her stimulating interpretation of this section, Maund points out the ambivalence of Hume's position:

despite the fact that Hume saw clearly that whatever an abstract idea is, it can-

[34] For Hume, Locke, and Berkeley on abstract ideas see the classic study of A. Meinong, 'Hume-Studien I: Zur Geschichte und Kritik des modernen Nominalismus', *Sitzungsberichte der philosophisch-historischen Klasse der Kaiserlichen Akademie der Wissenschaften in Wien* 87, 1877, pp. 185–260. See also Laird, *Hume*, pp. 55ff., and G. della Volpe, *La filosofia dell'esperienza di David Hume*, Rome 1972, pp. 54ff. An illuminating summary of the situation is in J. Yolton, *A Locke Dictionary*, Oxford, 1993, s.v. 'abstract or general ideas (and words)'.

not be an abstract image, he was still confused himself by his failure to distinguish ideas, in the widest sense of that word, and images.[35]

Kemp Smith's reading is similar: the central message of the section is that 'All Ideas are Images, and every Image is completely determinate'; 'The main Problem [whether abstract ideas are general or particular in the mind] is evaded, not solved'; and, on the whole, Hume's account of abstraction is 'confused'.[36]

Maund and Kemp Smith clearly have in mind mainly such passages as:

the mind cannot form any notion of quantity or quality without forming a precise notion of the degrees of each; (*T*/18)

the general idea of a line, notwithstanding all our abstractions and refinements, has in its appearance in the mind a precise degree of quantity and quality; however it may be made to represent others, which have different degrees of both; (*T*/19)

An idea is a weaker impression; and as a strong impression must necessarily have a determinate quantity and quality, the case must be the same with its copy or representative; (*T*/19)

The image in the mind is only that of a particular object, tho' the application of it in our reasoning be the same, as if it were universal. (*T*/20)

Their reading of these 'way-of-ideas' passages is unexceptionable: certainly in the *Treatise* abstract ideas are not general, but particular. It is also true that this raises the question: how is it that abstract ideas represent? What is not so clear is why one should follow Kemp Smith and Maund and consider such passages as more significant than the rest of the section.

The section on abstract ideas is approximately 9 pages long (*T*/17–25). Of these, 1 introduces the matter (*T*/17–18), 2 are devoted to show that ideas/images cannot be but particular (*T*/18–20), a bit more than 2 are on the role of resemblance and custom (*T*/20–2), then there are 2 pages on acts of the mind similar to that operating in the case of abstract ideas (*T*/22–4), and finally a bit more than 1 page is on the distinctions of reason (*T*/24–5). So, the identification of ideas and images is explicit in the first step of the argument (*T*/18–20); but it is only a small part of Hume's discussion of abstract ideas. The main theme of

35 Maund, *Knowledge*, p. 169.

36 The quotations are from three section titles in Kemp Smith, *Philosophy*, pp. 257, 261, 267. In pp. 267–68, Kemp Smith refers to Maund's book. See also Meinong and Laird (as in fn. 34).

the section is, I submit, not the relation or identification between ideas and images, but rather that between ideas and words. It may be useful to read this section more closely.

It is words that make individual ideas 'become general in their representation'. A general idea is a classification operated by the mind:

> When we have found a resemblance* among several objects, that often occur to us, we apply the same name to all of them, whatever differences we may observe in the degrees of their quantity and quality, and whatever other differences may appear among them. (*T*/20)

So, there is the frequent occurrence of several objects, the perception of the resemblance, and the application of the name, a sort of baptism where the decision is taken as to what resemblances and differences matter to our purposes. The asterisk after 'resemblance' refers to one of the 1740 additions, where it is made clear that we are not only talking of images:

> 'tis evident, that even different simple ideas may have a similarity or resemblance to each other . . . And of this we may be certain, even from the very abstract terms *simple idea*. (*T*/637)

The relation between words and ideas is complicated, and the main character of the section – the custom to call resembling objects with the same name – appears surrounded by resonances and associations:

> After we have acquired a custom of this kind, the hearing of that name revives the idea of one of those objects, and makes the imagination conceive it with all its particular circumstances and proportions. But as the same word is suppos'd to have been frequently applied to other individuals, that are different in many respects from that idea, which is immediately present to the mind; the word not being able to revive the idea of all these individuals, only touches the soul, if I may be allow'd so to speak, and revives that custom, which we have acquired by surveying them. (*T*/20)

The word 'only touches the soul' – words are evocative, custom is a sort of wizardry. The custom which operates here is 'one of the most extraordinary circumstances in the present affair' (*T*/21). It is 'so entire, that the very same idea may be annext to several different words, and may be employ'd in different reasonings, without any danger of mistake' (*T*/21).

The custom, 'which so readily recalls every particular idea, for which we may have occasion, and is excited by any word or sound, to which we commonly annex it' (*T*/22) is an act of the mind whose 'ultimate cause' is not known; but we can account for it by resorting to experience and analogy. For this custom is 'analogous' to other acts of the mind, such as, first:

> when we mention any great number, such as a thousand, the mind has generally no adequate idea of it, but only a power of producing such an idea, by its adequate idea of the decimals, under which the number is comprehended.

Secondly:

> we have several instances of habits, which may be reviv'd by one single word; as when a person, who has by rote any periods of a discourse, or any number of verses, will be put in remembrance of the whole, which he is at a loss to recollect, by that single word or expression, with which they begin. (*T*/23)

Thirdly, we do use terms to which 'we do not annex distinct and compleat ideas', and yet when we talk of 'government' and 'church' we 'may avoid talking nonsense' (*T*/23). Hume concludes with a fourth act of the mind, the one by which 'the fancy runs from one end of the universe to the other in collecting those ideas, which belong to any subject', as if there was operating 'a kind of magical faculty in the soul' (*T*/24). In short, the distinction between genuine ideas, the ones equipped with a correspondent image, and words which tend to mislead us, such as 'vacuum', is not clear-cut.

So, when we talk about a vacuum we (literally) do not know what we are talking about: we merely use words, and we take it for granted that they stand for ideas. In fact, we only have what we may call a pseudo-idea of a vacuum, which is the result not of our experience in the ordinary sense, but of our use of language. But then, what matters about the vacuum in this connection, and also what makes Hume's long discussion worthwhile, is not to know whether it exists or not, either within or without our mind, but rather that there is a way we commonly talk about it. At this point, however, it is worth noting that something starts to assume a new significance: the satirical mood Hume displays in this section. After having introduced the tendency to confuse ideas closely related, Hume writes:

> We might produce the figures of poets and orators, as sufficient proofs of this, were it as usual, as it is reasonable, in metaphysical subjects to draw arguments from that quarter. But lest metaphysicians shou'd esteem this below their dignity, I shall borrow a proof from an observation, which may be made on most of their own discourses, *viz*. that 'tis usual for men to use words for ideas, and to talk instead of thinking in their reasonings. (*T*/61–2)

This passage implies a whole style of thought: the love for Virgil and Cicero, the creative appropriation of the typical Port-Royal theme of the confusion between words and ideas, the spiteful poke at the arrogance and pomposity of metaphysicians, and their reduction to the state of ordinary human beings – even their using words for ideas is simply the result of a natural associative tendency. This brings us to the most important aspect of this whole discussion, an aspect which, curiously, appears only in the margins: abstract ideas are conversational, they imply both my own self and other people's. This is a crucial implication of the custom at the basis of the operation of general terms and abstract ideas: we use them in 'reflexion or conversation' (*T*/23–4); and we 'reason', 'mention', 'assert', 'make use of terms', 'talk', 'say' – all things which tend to happen in company. General terms and abstract ideas are intrinsically social.

NATURAL PHILOSOPHY AND THE 'SCIENCE OF HUMAN NATURE'

At the beginning of this chapter we saw that Hume's treatment of the conceivability of empty space sounds so alien to us because its terms of reference are in a standard Lockeanism with which we are no longer acquainted. A look at this tradition revealed that one of the reasons for the lack of balance of this section of the *Treatise* is the disappearance of God from a set of arguments where He had traditionally played a major role. Then, I moved to one of the main points of philosophical interest of the whole story: the elaboration of the notion of distance. In this way I came back to the 'manner of appearance' as the contribution of the self to sense perception. This, in turn, led me to focus on the problem of identifying simple visual impressions, then to consider abstract ideas and to try to sort out what it is that we are talking about when we use general terms. It is now time to answer the most basic question in this chapter, the question that has been left in the background even though it felt urgent from the start: why does Hume bother to discuss a non-existent idea, such as that of a vacuum? And how can he do so? Or, to put it in more reflexive terms: is it possible to read this section of the *Treatise* in a more interesting way than just saying that it is a misconceived set of pages written on automatic pilot? In my opinion, in order to answer this final question we must start by concentrating briefly on the science of human nature and on its distinctive features.

A good way to examine the features of the science of human nature is by contrasting them with natural philosophy. The section I am discussing is rich in materials on this score: the differences between natural philosophy and the science of human nature form a constant theme throughout the discussion of the conceivability of empty space. In particular, the refusal to engage in a natural philosophical discussion is explicit and repeated. For instance, Hume says of the third objection (the one according to which vacuum is necessary in order to make motion possible and conceivable):

> I shall not enlarge upon this objection, because it principally belongs to natural philosophy, which lies without our present sphere. (*T*/55)

Clearly natural philosophy is not of his concern.

Again, one finds natural philosophy – '*the Newtonian* philosophy', to be precise – in the addition which presents the second paradox (the one about 'touch'). The Newtonian philosophy is, according to Hume, a way of describing moving bodies, and has no further implications; to it 'Nothing is more suitable . . . than a modest scepticism to a certain degree, and a fair confession of ignorance in subjects, that exceed all human capacity' (*T*/639).[37] So, the distinction between the Newtonian philosophy and the 'science of human nature' is clear enough: the former teaches us how to talk about such subjects as moving and resting bodies, while the 'science of human nature' talks about the ways people are – about human ideas, errors, beliefs, delusions, passions. But then no space seems to be left for the discourse on infinite spaces and eternal times, the world and God – all subjects which would be out of place both in a natural philosophy thus conceived, and in the 'science of human nature'. God's absence and the consequent apparent lack of balance in Hume's discussion of empty space – one may at this point suggest – depend on the shift of focus from natural philosophy to human nature and on a drastic concomitant change of metaphysical mood. In order to characterise the science of human nature in its own right, it is necessary to find out what exactly this new, different metaphysical mood consists in. Before doing so, however, I think it may be interesting to consider the relationship between metaphysics and natural philosophy from a different point of view.

Space and time appear at the beginning of Hume's treatment of the

[37] It is interesting to compare this statement with the character of Newton that appears in the *History of England*: 'While Newton seemed to draw off the veil from some of the mysteries of nature, he showed at the same time the imperfections of mechanical philosophy: and thereby restored the ultimate secrets to that obscurity, in which they ever did and ever will remain' (ed. W. B. Todd, Indianapolis 1983, vol. 6, p. 542).

idea of cause and effect, before the introduction of the idea of necessary connection:

> we may ... consider the relation of *contiguity* as essential to that of causation ... if one cause were co-temporary with its effect, and this effect with *its* effect, and so on, 'tis plain there wou'd be no such thing as succession, and all objects must be co-existent. (*T*/75)

Causes and effects are conceived as contiguous and successive. Space and time, as they are introduced here, appear closer to the spirit of natural philosophy than anywhere else in the *Treatise* – they have, as it were, a less subjective air: the idea of cause and effect applies when, say, two billiard-balls appear in certain (spatial and temporal) mutual relations – in a certain 'manner', if one wants to keep Hume's expression. And here space is introduced as the contiguity, time as the succession of causes and effects. This is all very attractive, much more stimulating than the clumsy part of the *Treatise* which is the object of the present study. Also it is more pleasing than that conception of the ideas of space and time as derived from the never defined 'manner of appearance' of perceptions, or (even more confusedly) of elementary perceptual entities such as coloured and tangible points and notes played on a flute ... and yet, as Selby-Bigge observed, without the treatment of the ideas of space and time the analysis of causation is left without a foundation.[38] The whole thing holds together because there is also that derivation (confused and unclear, but at the same time independent and persuasive) of the ideas of space and time from the manner of appearance of common objects and of perceptual units. This derivation is the object of metaphysics, so that the relation between natural philosophy and metaphysics becomes a relation of dependence of the first on the second.[39] I can now go back to metaphysics, that is, to the 'science of human nature'.

I have said that metaphysics/the science of human nature talks about itself. It is also different from natural philosophy because it questions what natural philosophy assumes without further consideration. In doing so, natural philosophy operates correctly – questioning its own assumptions is none of its business. Metaphysics is bound to tackle

[38] Actually, Selby-Bigge says that it 'is left hanging in the air'. At any rate, he also says that 'the subject of space and time, as treated in the Treatise, is not very attractive'. See the introduction to his edition of the *Enquiries*, Oxford 1972 (2nd edn), p. xiii.

[39] By the way, this is exactly what Hume has promised in the introduction to the *Treatise*: 'Even Mathematics, Natural Philosophy, and Natural Religion, are in some measure dependent on the science of MAN; since they lie under the cognizance of men, and are judged of by their powers and faculties' (*T*/xv).

such questions, because it is by nature a self-questioning activity. The reason for this is that the task of the science of human nature, the true metaphysics is – as has been brilliantly put – 'to apprehend false reason and denounce it . . . the endless, Sisyphean labour . . . of putting an end to metaphysics'.[40] It is the never-ending task of defending people – with their human ideas, beliefs, natural errors, and delusions – from the perverse results of the natural, but misleading metaphysical instinct. The true metaphysicians, the confused metaphysicians responsible for false and adulterated metaphysics, and common people all live in the same philosophical and linguistic world; in fact, each full human being is a mix of true metaphysics, false metaphysics, and ordinary common sense, and all appear in the *Treatise*. To this situation is applicable what John Richetti has written about another page of the *Treatise*: 'I think [this passage] is comic, human in the Humean sense that the right to live with contradictions is earned by active inquiry'.[41]

Let us reconsider, for the last time, the paradox Hume presents in the appendix. It is worth quoting it once again:

> if it be ask'd, whether two objects, having such a distance betwixt them, touch or not: It may be answer'd, that this depends upon the definition of the word, *touch*. If objects be said to touch, when there is nothing *sensible* interpos'd betwixt them, these objects touch: If objects be said to touch, when their *images* strike contiguous parts of the eye, and when the hand *feels* both objects successively, without any interpos'd motion, these objects do not touch. (*T*/638–9)

Several pages ago, I observed in passing that this page has a Pyrrhonian flavour. Now I find that it is difficult not to think that Hume is answering Locke directly – in particular the passage in which Locke wonders, in a perfect analytic mood:

> Those who contend that Space and Body are the same, bring this Dilemma. Either this Space is something or nothing; if nothing be between two Bodies, they must necessarily touch; if it be allowed to be something, they ask, whether it be Body or Spirit? (II.xiii.sect 16)

But this time I would also like to try to make sense of it at a different level, by spelling out some of its implications. According to Hume, we know from the beginning that our way of talking about the world is not true-to-the-facts anyway; in fact, it is not even coherent, for words not only do not mirror reality, they do not mirror our ideas either. By

40 J. Rée, *Philosophical Tales*, London & New York 1987, p. 47.

41 J. Richetti, *Philosophical Writing*, Cambridge (Mass.) 1983, p. 221. Here Richetti is commenting on Hume's demonstration of the non-existence of the idea of the self in *T*/252.

deciding how to use the word 'touch', we decide how to play the game: we fix a way of talking about the world and choose what discourses are to be forbidden. There is nothing else to be known either about the world or about our image of the world. But there is still something very important to be known about us: the way we use words, our special way of exploiting language. We project onto language our insidious tendency towards consistency; we use words to fill the gaps in our ideas, to cover up the inevitable ambiguities in our thoughts.[42]

As a recapitulation and a conclusion of this chapter, I would like to quote from a section in Part 4, 'Of the ancient philosophy':

> as nature seems to have observ'd a kind of justice and compensation in every thing, she has not neglected philosophers more than the rest of the creation; but has reserv'd them a consolation amid all their disappointments and afflictions. This consolation consists in their invention of the words *faculty* and *occult quality*. For it being usual, after the frequent use of terms, which are really significant and intelligible, to omit the idea, which we wou'd express by them, and to preserve only the custom, by which we recall the idea at pleasure; so it naturally happens, that after the frequent use of terms, which are wholly insignificant and unintelligible, we fancy them to be on the same footing with the precedent, and to have a secret meaning, which we might discover by reflection. The resemblance of their appearances deceives the mind, as is usual, and makes us imagine a thorough resemblance and conformity. By these means these philosophers set themselves at ease, and arrive at last, by an illusion, at the same indifference, which the people attain by their stupidity, and true philosophers by their moderate scepticism. They only need say, that any phænomenon, which puzzles them, arises from a faculty or an occult quality, and there is an end of all dispute and enquiry upon the matter. (*T*/224)

Many things need to be said about this passage. First of all, the satirical attack on the philosophers is clearly of the same kind as the one in the discussion of the conceivability of a vacuum. I suggest that one should take it as a sign that Hume is performing one of his disruptive sceptical attacks. Second, Hume makes it rather clear that the mistake of the philosophers is in their attempt to distinguish themselves from the others – the sin of the snob. They take words for ideas as does anybody else, but are not prepared to admit it. Third, human beings take words for ideas; even philosophers do. The true philosopher takes words for ideas as well, and this is the reason for his moderate scepticism, for his

[42] Cf. *ibid.*, p. 219 (about *T*/262 and T. Penelhum's comments on it in his 'Hume on personal identity', in V. C. Chappell (ed.), *Hume: A Collection of Critical Essays*, Notre Dame, Ind., 1968), where more is credited to language and less to humans: 'Hume, it seems to me, is worried by just this predisposition of language to solve problems'.

serene and confident lack of self-confidence. But the reason why Hume is discussing the idea of empty space, which leads him to this moderately sceptical position, is that he himself, by dint of using the word, is the victim of a natural, spontaneous metaphysical drive and believes in empty space. The true philosophy, in other words, not only coexists with, it is actually parasitic on, the false one, it is the refutation of the false one. And what I find specially important and moving is that, while discussing the conceivability of empty space, the true philosopher is not sitting in his study as usual; for the whole discussion presupposes a community of talkative companions. So, he has got out and is actually talking and reasoning with his fellow human beings. Together they talk and talk (perhaps Hume is ready for his job as the ambassador between the learned and the conversible) and either retire into the obscure regions of vacua, occult qualities, and faculties, or denounce them – not as empty, but as fictitious. This is why the true philosopher has good reasons to take the discussions about the idea of a vacuum very seriously. So, I suggest, this is also why the fact that we reason and talk about extension means that we have an idea of it (since there is no reason to doubt), while talking about a vacuum does not mean that we have an idea of a vacuum. But this, in turn, does not mean that the 'idea of a vacuum' is not a good subject for a true philosopher to discuss. We can discuss the idea of empty space, even if we cannot conceive any such idea, because the erroneous conviction of conceiving an idea is at least as interesting as a true idea. When the true philosopher talks about the idea of a vacuum he knows perfectly well what he is talking about: he is talking about human nature.

CONCLUSION:

Space and the self

In Book 1, Part 2, sections 1–3 of the *Treatise*, Hume presents his doctrine of space. The first part of my study has been devoted to his rejection of the theory of its infinite divisibility and to his inquiry into the origin of the idea of space. In sections 4–5, Hume answers objections to his doctrine; and the second part of my study has focussed on the discussion to be found there of the nature of geometry and demonstrative knowledge, and of the idea of empty space. My overall aim has been to show that Hume's treatment of the idea of space is tightly linked with other, and for us more central, themes in the *Treatise*: the question of external objects, the problem of the unity of the self, the relation between knowledge and belief, and abstract ideas and distinctions of reason. I have argued that to understand Hume's sections on the idea of space we have to read them in the light of these other central issues in the *Treatise*, and, conversely, that the treatment of the idea of space may itself be used to cast light on these other issues, and thus on the philosophical project of the *Treatise* as a whole.

We have seen that the idea of space has something in common with the notion of belief: they are both 'manners' in which perceptions appear or are conceived. Both seem to be peculiar hybrids: they certainly cannot be ideas without correspondence with impressions, they do not seem to derive from any distinct impression of the senses, they do not exactly derive from passions or impressions of reflection either. This is why I have suggested that they may be regarded as traces, among mental contents, of mental operations and activities. Similarly, I have suggested that certainty depends on a reflexive quality of a particular kind of acts of the mind, and that another such reflexive quality is guiding the formation of our pseudo-idea of an independent space and of a vacuum. So Hume's sections on the idea of space may be regarded as investigations into the 'powers and qualities' of the mind. Hume regards metaphysics as a 'science of human nature', and the 'experience' constituting its object includes both 'impressions' and 'manners of appearance', both 'ideas' and 'manners of conception'; perceptions are its only objects, and the world and the mind are its two poles.

While the connection between space and the self is merely implicit in Book 1 of the *Treatise*, it is made fully explicit in Book 2, devoted to the passions. In chapter 2, I have mentioned Hume's double talk about the self.[1] In Book 1, in the section 'Of personal identity', the self is pulverised into a collection of independent experiences, a 'bundle' or 'connected heap' of perceptions, from which it is impossible ever to single it out – a result that Hume, in the Appendix to Book 3, declares unsatisfactory, and a disappointment of his own

> hopes, that however deficient our theory of the intellectual world might be, it wou'd be free from those contradictions, and absurdities, which seem to attend every explication, that human reason can give of the material world. (*T*/633)

But, in Book 2, when talking about such passions as pride and humility, or the love of fame, Hume says that our self is

> always so intimately present with us, and . . . our consciousness gives us so lively a conception of our own person, that 'tis not possible to imagine, that any thing can in this particular go beyond it. (*T*/317)

Many similar examples are possible: our 'intimate memory and consciousness' of self (*T*/277), our being 'intimately conscious' (*T*/286, 339) or 'every moment conscious' (*T*/340) of it, its being 'intimately present' to us (*T*/317, 320, 354, 427) appear over and over again in Book 2 of the *Treatise*. In this different frame of reference, the discussion of the relation between space and the self is open, and takes the form of an account of the concrete interactions constituting the spatial experience of consciousness.

> There is an easy reason, why every thing contiguous to us, either in space or time, shou'd be conceiv'd with a peculiar force and vivacity, and excel every other object, in its influence on the imagination. Ourself is intimately present to us, and whatever is related to self must partake of that quality. (*T*/427)

This is how Hume introduces the effects of contiguity or distance in space and time on passions. And then he proceeds to analyse:

> 'Tis obvious, that the imagination can never totally forget the points of space and time, in which we are existent; but receives such frequent advertisements

[1] See above, chapter 2, 'Manners of appearance and manners of conception'.

of them from the passions and the senses, that however it may turn its attention to foreign and remote objects, it is necessitated every moment to reflect on the present. 'Tis also remarkable, that in the conception of those objects, which we regard as real and existent, we take them in their proper order and situation, and never leap from one object to another, which is distant from it, without running over, at least in a cursory manner, all those objects, which are interpos'd betwixt them. (*T*/427–8)

The same applies to the distance between an object and ourselves:

When we reflect, therefore, on any object distant from ourselves, we are oblig'd not only to reach it at first by passing thro' all intermediate space betwixt ourselves and the object, but also to renew our progress every moment; being every moment recall'd to the consideration of ourselves and our present situation. (*T*/428)

Hume's particular point here is that this renewed passage step by step through the distance between ourselves and the distant object 'breaks' 'the action of the mind' through which the object is conceived, thus diminishing the vivacity of our conception; and, conversely, the fewer and smoother the steps from ourselves to the distant object, the less serious the weakening of the idea. Since it is considered with special regard to space, the self is seen (primarily) as one's body. And it is interesting that in this context, in connection with space and the passions, the already familiar way of accounting for the origin of the idea of space appears again, but with a new emphasis on the coexistence of the parts of spatial perceptions:

Without having recourse to metaphysics, any one may easily observe, that space or extension consists of a number of coexistent parts dispos'd in a certain order, and capable of being at once present to the sight or feeling. (*T*/429)

The relation between the self and the body marks yet again a difference between Books 1, and 2 of the *Treatise*. In Book 1 we find that our ability to identify our body as such, which appears as a presupposition of the centrality of the body in the passage quoted above, is, in fact, far from unmediated from the point of view of the understanding. We may take it for granted that 'our body evidently belongs to us'; but, in fact, from the point of view expressed in Book 1:

'Tis not our body we perceive, when we regard our limbs and members, but certain impressions, which enter by the senses; so that the ascribing a real and corporeal existence to these impressions, or to their objects, is an act of the mind as difficult to explain, as that [of the external existence of our perceptions]. (*T*/190–1)

Here the elements of spatial experience appear on full display: the central role of the body, the coexistent objects of sight and touch, their arrangement in the act of visual or tactile perception, the emotional effect of space in concrete experience. Our body itself, however, disappears as such, like the table in the section on the origin of the idea of space. It is dissolved, analysed into the impressions arising in our consciousness.

In Book 2, the situation is entirely different:

> Bodily pains and pleasures are the source of many passions, both when felt and consider'd by the mind; but arise originally in the soul, or in the body, whichever you please to call it, without any preceding thought or perception. A fit of the gout produces a long train of passions, as grief, hope, fear; but is not deriv'd immediately from any affection or idea. (*T*/276)

The self Hume is discussing here is the commonsensical self, immediately and unproblematically composed of body and mind, with which we are well acquainted from our ordinary life: 'pride and humility have the qualities of our mind and body, that is *self*, for their natural and more immediate causes' (*T*/303).[2] In the passage cited above about the effects of contiguity or distance on passions, reflection on any object 'distant from ourselves' requires that we renew every moment the passage 'thro' all intermediate space betwixt ourselves and the object', since we are 'every moment recall'd to the consideration of ourselves and our present situation' (*T*/428): here the body is the key to the distinction, expressed through the idea of space, between what is internal and what is external to the self; it is the constant standard in spatial perception, and thus, as it were, the reference point of consciousness. The link between the idea of space and the body of the perceiver is seen as one of coexistence, and affords the reason why 'the consequences of a removal in *space* are much inferior to those of a removal in *time*' (*T*/429): for,

> the parts of extension being susceptible of an union to the senses, acquire an union in the fancy; and as the appearance of one part excludes not another, the transition or passage of the thought thro' the contiguous parts is by that means render'd more smooth and easy. (*T*/429)

The relationship between space and time is thus established in terms of

[2] On the 'I' composed of body and soul, and on the perspective of common sense on it, see the inspiring remarks in N. Capaldi, 'The historical and philosophical significance of Hume's theory of the self', in A. J. Holland (ed.), *Philosophy, Its History and Historiography*, Dordrecht 1985, pp. 271–85, esp. pp. 280–1.

that between feeling and thinking, body and mind: and it is the link between space and the bodily self that makes our passions less immediately sensitive to spatial than to temporal distance. The main character of the page is a 'we' that, under an appearance of reassuringly unphilosophical normality and chattiness, is, in fact, completely anonymous, a self without a personal past, without any individual memories – a true bundle of perceptions.[3] The 'present' is the existence of this self, and is identified as 'the points of space and time, in which we are existent'.

I agree with all the Humean readers and interpreters who have found it remarkable that Hume does not even attempt to explain or excuse, let alone bridge, this gap between Books 1 and 2. But, as with a number of other problems left open in the *Treatise*, this is not, I think, a sign of inconsistency, or disengagement, or absent-mindedness: it is the apposite expression of a limit he has discovered, and of his whole conception of metaphysics itself as a 'science of human nature'. The difference between Books 1 and 2 is, in turn, the expression of a feature of human experience itself: consciousness is either projected in the isolation of a metaphysical emptiness where nothing can be taken for granted, or immersed in the atmosphere of our basic everyday objects, experiences, and familiarities and of our common human conversation. The 'science of human nature' is, in the end, both a history and a mirror of the human condition itself – a history of these alternatives and of the delicate balancing between them.

[3] On the anonymity of this self see G. Preti, *Alle origini dell'etica contemporanea. Adamo Smith*, Florence 1957, pp. 43, 57ff., and M. Dal Pra, 'L'astrazione nella filosofia di Hume', *Rivista Critica di Storia della Filosofia* 22 (4), 1967, pp. 435–56.

Bibliography

(In addition to works specifically cited in footnotes, I have included a number from which I have drawn inspiration or general guidance)

HUME'S WORKS AND CORRESPONDENCE

Birkbeck Hill, G. (ed.), *Letters of David Hume to William Strahan*, Oxford: Clarendon Press, 1888

Burton, J. H. *Life and Correspondence of David Hume*, 2 vols., Edinburgh: William Tait . . ., 1846 (repr. New York and London: Garland, 1983)

Green, T. H. and Grose T. H. (eds.), *The Philosophical Works of David Hume*, 4 vols., London: Longmans, Green, and Co., 1874–5

Greig, J. Y. T. (ed.), *The Letters of David Hume*, 2 vols., Oxford: Clarendon Press, 1932

Kemp Smith, N. (ed.), *Hume's Dialogues concerning Natural Religion*, 2nd edition with supplement, Edinburgh and London: Thomas Nelson and Sons, 1947 (1st edn Oxford 1935)

Keynes, J. M. and Sraffa, P. (eds.), *An Abstract of a Treatise of Human Nature, 1740. A Pamphlet hitherto Unknown of David Hume*. Reprinted with an introduction by J. M. Keynes and P. Sraffa, Cambridge University Press, 1938 (repr. Hamden, Connecticut: Archon Books, 1965)

Klibansky, R. and Mossner, E. C. *New Letters of David Hume*, Oxford: Clarendon Press, 1954

Miller, E. F. (ed.), *David Hume, Essays Moral, Political, and Literary*, revised edition, Indianapolis: The Liberty Classics, 1987

Selby-Bigge, L. A. (ed.), *Hume's Treatise of Human Nature*, ed. with analytical index by L. A. Selby-Bigge, 2nd edition with text revised and variant readings by P. H. Nidditch, Oxford: Clarendon Press, 1978

Hume's Enquiries concering the Human Understanding and concerning the Principles of Morals, ed. with introduction, comparative table of contents, and analytical index by L. A. Selby-Bigge, 3rd edition, with text revised and notes by P. H. Nidditch, Oxford: At the Clarendon Press, 1975

Todd, W. B. (ed.), *The History of England by David Hume*, 6 vols., Indianapolis: The Liberty Classics, 1983

CONTEMPORARY SOURCES

Anon., Review of Malezieu's *Elemens de geometrie . . .*, *Memoires de Trévoux*, September 1705, Article 129, pp. 1467–79

Anon., Review of the *Treatise of Human Nature*, *The History of the Works of the Learned*, November 1739, Article XXVI, pp. 353–404

Anon., Letter to *The Commonsense: The Englishman's Journal*, no. 178, 5 July 1740

Anon., Review of the *Treatise of Human Nature*, *Bibliothèque Raisonnée*, 1740:2, Article V, pp. 324–55

Anon., Review of T. Morgan, *Physico-Theology*, *The History of the Works of the Learned*, September 1741, Article XV, pp. 203–34, and October 1741, Article XVI, pp. 235–57

Arnauld, A. and Nicole, P. *La Logique ou l'Art de Penser*, P. Clair and F. Girbal eds., Paris: Presses Universitaires de France, 1965

Logic; or, the Art of Thinking: in which, Besides the 'Common', are contain'd many excellent 'New Rules', very profitable for directing of 'Reason', and acquiring of 'Judgment', in things as well relating to the Instruction of a 'Mans self', as of 'others'. In Four Parts . . . London: . . . Printed for H. Sawbridge . . . 1685

The Port-Royal Logic, translated from the French; with introduction, notes, and appendix, by T. S. Baynes, Edinburgh and London: William Blackwood and Sons, 1851

The Art of Thinking. Port-Royal Logic, transl. by J. Dickoff and P. James, with a foreword by C. W. Hendel, Indianapolis: The Library of Liberal Arts, Published by The Bobbs-Merrill Company Inc., 1964

Barrow, I. *The Usefulness of Mathematical Learning Explained and Demonstrated. Being Mathematical Lectures Read in the Public Schools at the University of Cambridge* by Isaac Barrow, D. D. Professor of Mathematics, and Master of the Trinity College, etc . . . . Translated by the Rev.d Mr. John Kirkby, of Egremont in Cumberland, London: Printed for Stephen Austen . . ., 1733 (reprt. London: Frank Cass & Co., 1970)

Baxter, A. *An Enquiry into the Nature of the Human Soul; Wherein the Immateriality of the Soul Is evinced from the Principles of Reason and Philosophy*, 2nd edition, London: Printed for the Author: And sold by A. Millar . . ., 1737 (1st edn 1733)

Bayle, P. *An Historical and Critical Dictionary . . . Translated into English, with many Additions and Corrections made by the Author himself, that are not in the French Editions*, 4 vols., London: for C. Harper, D. Brown, J. Tonson, A. and J. Churchill, T. Horne, T. Goodwin, R. Knaplock, J. Taylor, A. Bell, B. Tooke, D. Midwinter, B. Lintott, and W. Lewis, 1710

A General Dictionary, Historical and Critical; in which a new and accurate translation of that of . . . Mr Bayle . . . is included, and interspersed with several thousand lives never before published . . . With reflections on such passages of Mr Bayle, as seem to favour scepticism and the Manichee system. By the Rev. J. P. Bernard . . . Rev. Thomas Birch . . . John Lockman, and other hands, and the articles relating to Oriental history by George Sale, 10 vols., London: for G. Strahan, &c., 1734–41

Oeuvres diverses de Mr. Pierre Bayle, . . . contenant tout ce que cet Auteur a publié sur des matieres de Theologie, de Philosophie, de Critique, d'Histoire, et de Literature; excepté son Dictionnaire historique et critique, 4 vols., The Hague 1727–31;

vol. 4, chez P. Husson, T. Johnson, P. Gosse, J. Swart, H. Scheurleer, J. van Duren, C. le Vier, F. Boucquet; A Rotterdam, chez T. Johnson, 1731 (reprt. Hildesheim: Georg Holms Verlagsbuchhandlung, 1964–70)

Berkeley, G. *The Works of George Berkeley, Bishop of Cloyne*, ed. by A. A. Luce and T. E. Jessop, 9 vols., London: Thomas Nelson & Sons, 1948–57

Bernard, J. Review of Malezieu's *Elemens de geometrie* . . ., *Les Nouvelles de la Republique des Lettres*, September 1705, pp. 340–57

Boswell, J. *Boswell Laird of Auchinleck, 1778–1782*, ed. by J. W. Reed and F. A. Pottle, New York, Toronto, and London: McGraw-Hill Book Company, 1977

Life of Samuel Johnson, Oxford University Press, 1980

Browne, P. *The Procedure, Extent and Limits of Human Understanding*, London: Printed for William Innys . . ., 1728

Things Divine and Supernatural conceived by Analogy with Things Natural and Human. By the Author of the Procedure . . ., London: Printed for William Innys and Richard Manby . . ., 1733

Chambers, E. *Cyclopaedia: Or, an Universal Dictionary of Arts and Sciences* . . . by E. Chambers, FRS . . . in 2 volumes . . ., London 1738

Cheyne, G. *Philosophical Principles of Natural Religion: Containing the Elements of Natural Philosophy And the Proofs for Natural Religion, Arising from them*, London: Printed for George Strahan . . ., 1705

Clarke, S. *A Demonstration of the Being and Attributes of God. More Particularly in Answer to Mr. Hobbs, Spinoza, and their Followers. Wherein the Notion of Liberty is Stated, and the Possibility and Certainty of it Proved, in Opposition to Necessity and Fate. Being the Substance of Eight Sermons, Preached in the Cathedral-Church of St Paul, in the Year 1704, at the Lecture Founded by the Honourable Robert Boyle, Esq;* . . ., London: Printed by Will. Botham, for James Knapton . . ., 1705

Descartes, R. *Oeuvres de Descartes*, publiées par Charles Adam et Paul Tannery, nouvelle présentation, en co-édition avec le Centre National de la Recherche Scientifique, Paris: Librairie Philosophique J. Vrin, vol. 8, 1964, vol. 10, 1966

The Philosophical Works of Descartes, rendered into English by E. S. Haldane and G. R. T. Ross, 2 vols., Cambridge University Press, 1911–12

Dillon, W. Earl of Roscommon *Essay on Translated Verse*, London: Printed for J. Tonson, 1684

Green, R. *The Principles of Natural Philosophy. In which is shewn the Insufficiency of the Present Systems, to give us any Just Account of that Science: And the Necessity there is of some New Principles. In Order to furnish us with a True and Real Knowledge of Nature*, Cambridge: Printed at the University-Press, for Edm. Jeffery Bookseller in Cambridge . . ., 1712

Heineccius, J. G. *Elementa philosophiae rationalis et moralis, Ex Principiis admodum evidentibus justo ordine adornata. Praemissa est Historia Philosophica. Editio quinta accuratissima*, in *Opera*, 8 vols., vol. 1, Genevae: Impensis Hered. Cramer et Fratr. Philibert, 1744

Jackson, J. *The Existence and Unity of God, proved from his Nature and Attributes, being a Vindication of Dr. Clarke's Demonstration of the Being and Attributes of God* . . ., London: Printed for James, John, and Paul Knapton . . ., 1734

Johnson, S. *A Dictionary of the English Language: In which The Words are deduced from their Originals, and Illustrated in their Different Significations by Examples from the best Writers* . . ., 2 vols., London: Printed by W. Strahan, for J. and P. Knapton; T. and T. Longman; C. Hitch and L. Hawes; A. Millar; and R. and J. Dodsley, 1755 (reprt. Harlow: Longman, 1990)

Kant, I. *Critique of Pure Reason*, transl. N. Kemp Smith, London: Macmillan, 1985 (1st edn London 1929)

Prolegomena to Any Future Metaphysics, transl. by L. W. Beck, Indianapolis and New York: Bobbs-Merrill Company Inc., 1950

Critique of Practical Reason, transl. by L. W. Beck, Indianapolis: Bobbs-Merrill Educational Publishing, 1956

King, W. *De Origine Mali*, Londini: Impensis Benj. Tooke . . ., 1702

An Essay on the Origin of Evil. By Dr. William King, late Lord Archbishop of Dublin. Translated from the Latin, with Large Notes; tending to explain and vindicate some of the Author's Principles Against the Objections of Bayle, Leibnitz, the Author of a Philosophical Enquiry concerning Human Liberty A. Collins; and others. To which is prefix'd a Dissertation concerning the Fundamental Principle and Immediate Criterion of Virtue. As also the Obbligation to, and Approbation of it. With some account of *the Origin of the Passions and Affections.* London: Printed for W. Thurlbourn bookseller in Cambridge . . . 1731

An Essay on the Origin of Evil. By Dr. William King, late Lord Archbishop of Dublin. Translated from the Latin, with Notes; and a Dissertation concerning the Principles and Criterion of Virtue and the Origin of the Passions. The Second Edition corrected and enlarged from the Author's Manuscripts. To which are added two Sermons by the same Author, the former concerning Divine Prescience; the latter on the Fall of Man, never before publish'd. By Edmund Law, M.A. Fellow of Christ's College in Cambridge. London: Printed by J. Stephens, for W. Thurlbourn Bookseller in Cambridge, . . . 1732

A Sermon Preached at Christ-Church, Dublin, May 15. 1709 . . . Dublin: Printed by, and for J. Gowan . . . 1727 (repr. in A. Carpenter (ed.), *Archbishop King's Sermon on Predestination*, introduced by D. Berman, Dublin: The Cadenus Press, 1976)

Law, E. *An Enquiry into the Ideas of Space, Time, Immensity, and Eternity; as also the Self-Existence, Necessary Existence, and Unity of the Divine Nature: In Answer to a Book lately Publish'd by Mr. Jackson, entitled, The Existence and Unity of God proved from his Nature and Attributes.* By Edmund Law, M. A. Fellow of Christ-College in Cambridge . . . Cambridge: Printed by W. Fenner and R. Beresford, for W. Thurlbourn; . . . 1734 (repr. Bristol: Thoemmes, 1990)

Considerations on the State of the World with regard to the Theory of Religion, in Three parts . . ., Cambridge: Printed by J. Bentham Printer to the University, for W. Thurlbourn Bookseller in Cambridge, and J. Beecroft . . ., 1745

Leibniz, G. W. and Clarke, S. *Correspondence*, ed. by H. G. Alexander, Manchester University Press, 1956

Locke, J. *An Essay concerning Human Understanding*, ed. by P. H. Nidditch, Oxford: Clarendon Press, 1975

Malebranche, N. *De la Recherche de la vérité*, ed. by G. Rodis-Lewis, in A. Robinet et al. (eds.), *Oeuvres complètes*, vols. 1–2, Paris: CNRS, 1962

Father Malebranche's Treatise concerning The Search after Truth. The Whole Work compleat . . . All translated by T. Taylor, M.A. of Magd. Coll. in Oxford. Oxford, Printed . . . for Thomas Bennet Bookseller . . ., London, 1694

The Search After Truth, translated from the French by T. M. Lennon and P. J. Oscamp, Columbus: Ohio State University Press, 1980

Malezieu, N. *Elemens de Geometrie de Monseigneur le Duc de Bourgogne, Troisième edition, Revûë, corrigée & augmentée d'un Traité des Logarithmes, Par M. De Malezieu. Avec l'Introduction a l'Application de l'algebre a la geometrie*, A Paris: Chez Estienne Ganeau . . ., 1729

Martin, B. *The Philosophical Grammar; Being a View of the Present State of Experimental Physiology, or Natural Philosophy*, London: Printed for J. Noon, 1735

Nieuwentijtd, B. *The Religious Philosopher: or, the Right Use of Contemplating the Works of the Creator, . . . designed for the conviction of atheists and infidels . . . translated by J. Chamberlayne . . .*, London: Printed for J. Senex, 1718 (1st edn in Dutch, Amsterdam 1714)

Pascal, B. *Pensées*, ed. by Z. Tourneur and D. Anzieu, 2 vols., Paris: Librairie Armand Colin, 1960

Pope, A. *The Works of Alexander Pope, Esq . . . With his last corrections . . .etc., Printed Verbatim from the Octavo Edition of Mr. Warburton*, 10 vols., London: Printed for J. and P. Knapton, H. Lintot, J. and R. Tonson and S. Draper, and C. Bathurst, 1754

Ralphson, J. *A Mathematical Dictionary: Or, a Compendious Explication of All Mathematical Terms, Abridg'd from Monsieur Ozanam, and Others*, London: Printed for J. Nicholson at the King's Arms . . . 1702

Reid, T. *An Inquiry into the Human Mind, on the Principles of Common Sense*, 4th edition, London: Printed for T. Cadell . . . London; and J. Bell and W. Creech, Edinburgh, 1785 (repr. Bristol: Thoemmes, and Tokyo: Kinokuniya, 1990)

Sterne, L. *The Life and Opinions of Tristram Shandy*, M. New, J. New, R. A. Davies, and W. G. Day ed. by, Gainesville: University Presses of Florida, 1984, 3 vols.

Whately, R. *Historic Doubts Relative to Napoleon Bonaparte*, London: Printed for J. Hatchard . . ., 1819

Whewell, W. (ed.), *The Mathematical Works of Isaac Barrow D. D.*, Cambridge 1860

SECONDARY LITERATURE

Andersen, K. 'Cavalieri's method of indivisibles', *Archive for the History of the Exact Sciences* 31, 1984–5, pp. 291–367

Anderson, R. F. *Hume's First Principles*, Lincoln: University of Nebraska Press, 1966

'The location, extension, shape and size of Hume's perceptions', in D. W. Livingston and J. T. King (eds.), *Hume: A Re-evaluation*, New York: Fordham University Press, 1976, pp. 153–71

Anon., Review of Burton's *Life and Letters of David Hume*, in *Quarterly Review* 78, 1846, Art. VI, pp. 75–113

Atkinson, R. F. 'Hume on mathematics', *Philosophical Quarterly* 10, 1960, pp. 127–37

Ayer, A. J. *Language, Truth and Logic*, 2nd edition, London: Gollancz, 1946 (1st edn 1936)

Hume, Oxford University Press, 1980

Ayers, M. 'Berkeley and Hume: a question of influence', in R. Rorty, J. B. Schneewind, and Q. Skinner (eds.), *Philosophy in History: Essays on the Historiography of Philosophy*, Cambridge University Press, 1984, pp. 303–27

Baier, A. C. *A Progress of Sentiments. Reflections on Hume's Treatise*, Cambridge (Mass.) and London: Harvard University Press, 1991

Barfoot, M. 'Hume and the culture of science in the early eighteenth century', in M. A. Stewart (ed.), *Studies in the Philosophy of the Scottish Enlightenment*, Oxford: Clarendon Press, 1990, pp. 151–90

Baroncelli, F. *Un inquietante filosofo perbene: saggio su David Hume*, Florence: La Nuova Italia, 1975

Baxandall, M. *Patterns of Intention. On the Historical Explanation of Pictures*, New Haven and London: Yale University Press, 1985

Baxter, D. M. 'Hume on infinite divisibility', *History of Philosophy Quarterly* 5, 1988, pp. 133–40

Beauchamp, T. L., Biro, J. I. and McIntyre, J. L. 'A discussion on personal identity', *Hume Studies* 5 (1), 1979

Beck, L. W. *Essays on Kant and Hume*, New Haven: Yale University Press, 1978

Bennett, J. *Locke Berkeley Hume: Central Themes*, Oxford: Clarendon Press, 1971

Bermúdez, J. L., Marcel A., and Eilan N. (eds.), *The Body and the Self*, Cambridge (Mass.): MIT Press, 1995

Blackburn, S. *Essays in Quasi-Realism*, Oxford University Press, 1993

Blay, M. 'Deux moments de la critique du calcul infinitésimal: Michel Rolle et George Berkeley', *Revue d'Histoire des Sciences* 39 (3), 1986, pp. 223–53

Bloom, H. *The Western Canon: The Books and School of the Ages*, New York, San Diego, and London: Harcourt Brace & Company, 1994

Box, M. *The Suasive Art of David Hume*, Princeton University Press, 1990

Bracken, H. M. *The Early Reception of Berkeley's Immaterialism 1710–1733*, The Hague: Martinus Nijhoff, 1965 (2nd edn; 1st edn 1958)

Bricke, J. *Hume's Philosophy of Mind*, Princeton University Press, 1980

Broad, C. D. 'Hume's theory of space', *Proceedings of the British Academy* 47, 1961, pp. 161–76

Broome, J. 'On the authorship of the *Abstract*: a reply to John O. Nelson', *Hume Studies* 18 (1), 1992, pp. 95–104

Burnyeat, M. F. 'The sceptic in his place and time', in R. H. Popkin and C. B. Schmitt (eds.), *Scepticism from the Renaissance to the Enlightenment*, Wolfenbütteler Forschungen, 35, Wiesbaden: Otto Harrassowitz, 1987, pp. 13–43

Burtt, E. A. *The Metaphysical Foundations of Modern Science*, London: Routledge & Kegan Paul, 1980

Butler, R. J. '*Distinctio Rationis*, or the Cheshire Cat which left its smile behind it', *Proceedings of the Aristotelian Society*, N.S. 76, 1976, pp. 165–176

'Hume's impressions', in G. Vesey (ed.), *Impressions of Empiricism*, Royal Institute of Philosophy Lectures, vol. 9, 1974–75, London: Macmillan, 1976, pp. 122–36.

Cantor, G. *Optics after Newton. Theories of Light in Britain and Ireland 1704–1840*, Manchester University Press, 1983

'Berkeley's *The Analyst* revisited', *Isis* 74, 1984, pp. 668–83

Capaldi, N. 'The historical and philosophical significance of Hume's theory of the self', in A. J. Holland (ed.), *Philosophy, Its History and Historiography*, Dordrecht: Reidel, 1985, pp. 271–85

Cassam, Q. (ed.), *Self-Knowledge*, Oxford University Press, 1994

Cassirer, E. *Die Erkenntnisproblem in der Philosophie und Wissenschaft der neueren Zeit*, vol. 2, Berlin: Verlag von Bruno Cassirer, 1907

Christensen, J. *Practicing Enlightenment. Hume and the Formation of a Literary Career*, Madison and London: University of Wisconsin Press, 1987

Church, R. W. *Hume's Theory of the Understanding*, London: George Allen and Unwin Ltd., 1968 (1st edn 1935)

Coleman, D. 'Is mathematics for Hume synthetic a priori?', *Southwestern Journal of Philosophy* 10, 1979, pp. 113–26

Connon R. W. and Pollard, M. 'On the authorship of "Hume's" *Abstract*', *Philosophical Quarterly* 27, 1977, pp. 60–6

Craig, E. *David Hume. Eine Einführung in seine Philosophie*, Frankfurt-on-Main: Vittorio Klostermann, 1979

The Mind of God and the Works of Man, Oxford: Clarendon Press, 1987

Cummins, P. 'Bayle, Leibniz, Hume and Reid on extension, composites and simples', *History of Philosophy Quarterly* 7, 1990, pp. 299–314

Cupitt, D. *Christ and the Hiddenness of God*, London: SCM, 1971

Dal Pra, M. 'L'astrazione nella filosofia di Hume', *Rivista Critica di Storia della Filosofia* 22 (4), 1967, pp. 435–56

Daniels, N. *Thomas Reid's Inquiry. The Geometry of Visibles and The Case for Realism*, New York: Burt Franklin & Co., 1974

Davie, G. 'Edmund Husserl and "the as yet, in its most important respect, unrecognized greatness of Hume"', in G. P. Morice (ed.), *David Hume, Bicentenary Papers*, Edinburgh: At the University Press, 1977, pp. 69–76

A Passion for Ideas: Essays on the Scottish Enlightenment, vol. 2, Edinburgh: Polygon, 1994

Deleuze, G. *Empirisme et subjectivité. Essai sur la nature humaine selon Hume*, Paris: Presses Universitaires de France, 3rd edition, 1973 (1st edn 1953)

Della Volpe, G. *La filosofia dell'esperienza di David Hume*, Roma: Editori Riuniti, 1972
Di Stefano, M. E. and Frasca-Spada, M. 'Metodi classici e trattazione del continuo: alcuni aspetti della storia degli indivisibili', in C. Maccagni and P. Freguglia (eds.), *La storia delle scienze*, Rome: Bramante, 1989, pp. 158–92
Donnici, R. *Husserl e Hume: per una fenomenologia della natura umana*, Milan: Franco Angeli, 1989
Eagleton, T. *The Ideology of the Aesthetic*, Oxford: Blackwell, 1990
Echelbarger, C. 'Hume on mathematics', paper presented at the 22nd Annual Conference of the Hume Society, Park City, Utah, July 1995
Eliot, T. S. 'Tradition and the individual talent', *Selected Essays*, 3rd edition, London: Faber & Faber, 1951, p. 22
Evans, A. W. *Warburton and the Warburtonians: A Study in Some Eighteenth-Century Controversies*, London: Oxford University Press, 1932
Falkenstein, L. 'Hume's account of space and time as manners or dispositions', *Archiv für Geschichte der Philosophie* 79, 1997, pp. 179–201
Feingold M. (ed.), *Before Newton. The Life and Times of Isaac Barrow*, Cambridge University Press, 1990
Flage, D. *Hume's Theory of Mind*, London and New York: Routledge, 1990
'Hume's relative ideas', *Hume Studies* 7, 1981, pp. 55–73
'Relative ideas revisited', *Hume Studies* 8, 1982, pp. 158–71
Flew, A. *Hume's Philosophy of Belief. A Study of His First Inquiry*, London: Routledge & Kegan Paul, 1961
'Did Hume distinguish pure from applied geometry?', *Ratio* 8, 1966, pp. 96–100
'Infinite divisibility in Hume's *Treatise*', in D. W. Livingston and J. T. King (eds.), *Hume: A Re-evaluation*, New York: Fordham University Press, 1976, pp. 257–69 (First published in *Rivista Critica di Storia della Filosofia* 22, 1967, pp. 457–71)
David Hume Philosopher of Moral Science, Oxford: Basil Blackwell, 1986
Fogelin, R. *Hume's Scepticism in the Treatise of Human Nature*, London: Routledge & Kegan Paul, 1985
Philosophical Interpretations, New York and Oxford: Oxford University Press, 1992
Force, J. E. 'Hume's interest in Newton and science', *Hume Studies* 14, 1988, pp. 166–216
Franklin, J. 'Achievements and fallacies in Hume's account of infinite divisibility', *Hume Studies* 20 (1), 1994, pp. 85–101
Frasca-Spada, M. 'Hume sullo spazio vuoto', *Studi Settecenteschi* 7–8, 1985–6, pp. 191–204
'Some aspects of Hume's conception of space', *Studies in History and Philosophy of Science* 21 (3), 1990, pp. 371–411
'Quixotic confusions and Hume's imagination: the "science of human nature" on the beliefs and sentiments of readers' (forthcoming)
Furley, David J. *Two Studies in the Greek Atomists*, Princeton University Press, 1967

Garrett, D. *Cognition and Commitment in Hume's Philosophy*, Oxford University Press, 1997

Gascoigne, J. *Cambridge in the Age of Enlightenment: Science, Religion and Politics from the Restoration to the French Revolution*, Cambridge University Press, 1989

Giusti, E. *Bonaventura Cavalieri and the Theory of Indivisibles*, Rome: Gremese, 1980

'Dopo Cavalieri. La discussione sugli indivisibili', in L. Grugnetti and O. Montaldo (eds.), *La Storia delle Matematiche in Italia. Atti del Convegno 29 Settembre-1 Ottobre*, Cagliari: University of Cagliari, 1984, pp. 85–114

Golden, M. 'Periodical context in the imagined world of Tristram Shandy', in P. J. Korshin (ed.), *The Age of Johnson. A Scholarly Annual*, New York: AMS Press, 1987, pp. 237–60

Gossmann, L. 'Two unpublished essays on mathematics in the Hume papers', *Journal of the History of Ideas* 21, 1960, pp. 442–9

Gotterbarn, D. 'Kant, Hume and analyticity', *Kant-Studien* 65, 1974, pp. 274–83

Graham, W. *English Literary Periodicals*, New York: Nelson & Sons, 1930

Guicciardini, N. *The Development of the Newtonian Calculus in Britain 1700–1800*, Cambridge University Press, 1989

Hearn, T. K. Jr., '"General rules" in Hume's *Treatise*', *Journal of the History of Philosophy* 8, 1970, pp. 405–22

Hendel, C. W. *Studies in the Philosophy of David Hume*, Princeton University Press, 1925; 2nd edition, Indianapolis: Bobbs-Merrill, 1963

Henrich, D. 'Identität und Objektivität. Eine Untersuchung über Kants tranzendentale Deduktion', *Sitzungberichte der Heidelberger Akademie der Wissenschaften. Philosophische-historische Klasse*, 1976, pp. 1–112

Holtby, R. T. *Daniel Waterland 1683–1740. A Study in Eighteenth-Century Orthodoxy*, Carlisle: Charles Thurnam and Sons Ltd., 1966

Hunt, J. *Religious Thought in England from the Reformation to the End of Last Century. A Contribution to the History of Theology*, London: Strahan and Co., 1870–73

Hurlbutt, R. H. *Hume, Newton and the Design Argument*, Lincoln: University of Nebraska Press, 1965

Huxley, T. H. *Hume: with Helps to the Study of Berkeley*, London: Macmillan, 1897

Jacquette, D. 'Kant's second antinomy and Hume's theory of extensionless indivisibles', *Kant-Studien* 98, 1993, pp. 38–50

'Infinite divisibility in Hume's first *Enquiry*', *Hume Studies* 20, 1994, pp. 219–40

'Hume's aesthetic psychology of distance, greatness and the sublime', *British Journal for the History of Philosophy* 3, 1995, pp. 89–112

'Hume on infinite divisibility and sensible extensionless indivisibles', *Journal of the History of Philosophy* 34, 1996, pp. 61–78

Jessop, T. E. 'The misunderstood Hume', in W. B. Todd (ed.), *Hume and the Enlightenment*, Edinburgh University Press, 1974

Johnson, D. M. 'Hume's missing shade of blue, interpreted as involving habitual spectra', *Hume Studies* 10 (2), 1984, pp. 109–24

Jones, P. *Hume's Sentiments. Their Ciceronian and French Context*, Edinburgh University Press, 1982

Kemp Smith, N. *The Philosophy of David Hume*, London: Macmillan & Co., 1941
Kermode, F. *The Classic*, 2nd edition, Cambridge (Mass.): Harvard University Press, 1983
Kuhns, R. 'Hume's republic and the universe of Newton', in P. Gay (ed.), *Eighteenth Century Studies Presented to Arthur M. Wilson*, New York: Russell & Russell, 1972
Kuypers, M. S. *Studies in the Eighteenth Century Background of Hume's Empiricism*, Minneapolis 1930
Laing, B. M. *David Hume*, 2nd edition, New York 1968 (1st edn London: E. Benn, 1932)
Laird, J. *Hume's Philosophy of Human Nature*, London: Methuen & Co., 1932
Leroy, A.-L. *David Hume*, Paris: Presses Universitaires de France, 1953
Préface du traducteur, in Hume, *Traité de la nature humaine*, Traduction, Préface et Notes de André Leroy, Paris: Aubier, 1946
Lievers, M. 'The Molyneux Problem', *Journal of the History of Philosophy* 30, 1992, pp. 399–416
Livingston, D. W. *Hume's Philosophy of Common Life*, Chicago and London: University of Chicago Press, 1984
'A Sellarsian Hume?', *Journal of the History of Philosophy* 29 (2), 1991, pp. 281–90
Luce, A. A. *Berkeley and Malebranche. A Study in the Origins of Berkeley's Thought*, Oxford: Clarendon Press, 1934
MacCracken, C. J. *Malebranche and British Philosophy*, Oxford: Clarendon Press, 1983
Malherbe, M. *Hume*, Paris: Vrin, 1976
Mall, R. A. *Experience and Reason. The Phenomenology of Husserl and its Relation to Hume's Philosophy*, The Hague: Martin Nijhoff, 1973
Mancosu, P. 'The metaphysics of the calculus: a foundational debate in the Paris Academy of Sciences, 1700–1706', *Historia Mathematica* 16, 1989, pp. 224–48
Maund, C. *Hume's Theory of Knowledge*, London: Macmillan & Co., 1937
Meinong, A. 'Hume-Studien I: Zur Geschichte und Kritik des modernen Nominalismus', *Sitzungsberichte der philosophisch-historischen Klasse der Kaiserlichen Akademie der Wissenschaften in Wien*, 87, 1877, pp. 185–260
Michaud, A. 'Remarques sur l'atomisme et le phénoménalisme de Hume', *Les Etudes Philosophiques* 46, 1973, pp. 43–57
Mijuskovic, B. 'Hume on space (and time)', *Journal of the History of Philosophy* 6 (4), 1977, pp. 387–94
Mossner, E. C. *The Life of David Hume*, Edinburgh: Thomas Nelson & Sons, 1954
'Hume's epistle to Dr. Arbuthnot, 1734: biographical significance', *Huntington Library Quarterly* 7, 1943–4, pp. 135–52
'The continental reception of Hume's *Treatise*, 1730–1741', *Mind* 56, 1947, pp. 31–43
'Hume's early memoranda, 1729–1740: the complete text.', *Journal of the History of Ideas* 9, 1948, pp. 492–518

'The first answer to Hume's *Treatise*: an unnoticed item of 1740', *Journal of the History of Ideas*, 1951, pp. 291–4

Mullan, J. *Sentiment and Sociability. The Language of Feeling in the Eighteenth Century*, Oxford: Clarendon Press, 1988

Murphy, R. T. *Hume and Husserl. Towards Radical Subjectivism*, The Hague: Martin Nijhoff, 1980

Nelson, J. O. 'Hume's missing shade of blue reviewed', *Hume Studies* 15 (2), 1989, pp. 353–63

'Has the authorship of the *Abstract of a Treatise of Human Nature* really been decided?', *Philosophical Quarterly* 26, 1976, pp. 82–91

'The authorship of the *Abstract* revisited', *Hume Studies* 17 (1), 1991, pp. 83–6

Newman, R. 'Hume on space and geometry', *Hume Studies* 7 (1), 1981, pp. 1–31

Nicolson, M. H. *Science and Imagination*, Ithaca and New York: Cornell University Press, 1956

Nidditch, P. H. *An Apparatus of Variant Readings for Hume's Treatise of Human Nature*, Dept. of Philosophy, University of Sheffield 1976

Norton, D. F. 'More evidence that Hume wrote the *Abstract*', *Hume Studies* 19 (1), 1993, pp. 217–22

Noxon, J. *Hume's Philosophical Development. A Study of His Methods*, Oxford: Clarendon Press, 1973

Olson, R. *Scottish Philosophy and British Physics 1750–1880. A Study in the Foundations of the Victorian Scientific Style*, Princeton University Press, 1975

Pappas, G. S. 'Abstract general ideas in Hume', *Hume Studies* 15 (2), 1989, pp. 339–52

Parusnikova, Z. 'Against the spirit of foundations: post-modernism and David Hume', *Hume Studies* 19 (1), 1993, pp. 1–17

Passmore, J. *Hume's Intentions*, 3rd edition, London: Duckworth, 1980 (1st edn Cambridge 1952)

A Hundred Years of Philosophy, Harmondsworth: Penguin Books Ltd., 1968 (1st edn London: Duckworth, 1957)

Pears D. F. (ed.), *David Hume. A Symposium*, London: Macmillan & Co., 1963

Hume's System. An Examination of the First Book of his Treatise, Oxford University Press, 1990

Penelhum, T. 'Hume on personal identity', in V. C. Chappell (ed.), *Hume: A Collection of Critical Essays*, Notre Dame, Ind.: University of Notre Dame Press, 1968, pp. 213–39

Phillipson, N. *Hume*, London: Weidenfeld & Nicholson, 1989

Pittion, J. P. 'Hume's reading of Bayle: an inquiry into the source and role of the *Memoranda*', *Journal of the History of Philosophy* 15, 1977, pp. 373–86

Pocock, J. G. A. *Virtue, Commerce, and History: Essays on Political Thought and History, Chiefly in the Eighteenth Century*, Cambridge University Press, 1985

Popkin, R. 'The early critics of Hume', in *The High Road to Pyrrhonism*, San Diego: Austin Hill Press, 1980, pp. 197–212

Preti, G. *Alle origini dell'etica contemporanea. Adamo Smith*, Florence: La Nuova Italia, 1957

Price, J. V. *The Ironic Hume*, Austin: University of Texas Press, 1965 (repr. Bristol: Thoemmes, 1992)

Pritchard, H. A. *Knowledge and Perception*, Oxford: Clarendon Press, 1950

Raynor, D. '"*Minima sensibilia*" in Berkeley and Hume', *Dialogue* 19, 1980, pp. 196–9

'The authorship of the *Abstract* revisited', *Hume Studies* 19 (1), 1993, pp. 213–16

Rée, J. *Philosophical Tales. An Essay on Philosophy and Literature*, London and New York: Methuen, 1987

Reichenbach, H. *The Rise of Scientific Philosophy*, Berkeley and Los Angeles: University of California Press, 1951

Reinach, A. 'Kants Auffassung des Humeschen Problems', *Zeitschrift für Philosophie und philosophische Kritik* 141, 1908, pp. 176–209 (English transl. by J. N. Mohanty, 'Kant's interpretation of Hume's Problem', in K. R. Merrill and R. W. Shahan (eds.), *David Hume, Many-Sided Genius*, Norman: University of Oklahoma Press, 1976, pp. 161–88)

Richetti, J. J. *Philosophical Writing: Locke, Berkeley, Hume*, Cambridge (Mass.): Harvard University Press, 1983

Rivers, I. (ed.), *Books and their Readers in Eighteenth Century England*, New York: Leicester University Press and St Martin's Press, 1982

Rollin, B. 'Hume's blue patch', *Journal of the History of Ideas* 33, 1971, pp. 119–28

Rorty, R. 'The pragmatist's progress', in U. Eco, *Interpretation and Overinterpretation*, ed. by S. Collini, Cambridge University Press, 1992, pp. 89–108

Ryley, R. M. 'Did Warburton review Hume's *Treatise of Human Nature*?', *Notes and Queries* 221, 1976, 354–5

Salmon, C. V. 'The central problem of David Hume's philosophy: an essay toward a phenomenological interpretation of the first book of the *Treatise of Human Nature*', *Jahrbuch für Philosophie und Phänomenologische Forschung* 10, 1929, 299–449

Siebert, D. T. *The Moral Animus of David Hume*, London and Toronto: Associated University Presses, 1990

'"Ardour of youth": the manner of Hume's *Treatise*', in R. Ginsberg (ed.), *The Philosopher as a Writer: The Eighteenth Century*, London and Toronto: Associated University Presses, 1987

Sitter, J. E. *Literary Loneliness in Mid-Eighteenth Century England*, Ithaca and London: Cornell University Press, 1982

Skinner, Q. 'Meaning and understanding in the history of ideas', *History and Theory* 8, 1969, pp. 3–53

Spitz, J. F. 'Le Problème de l'analyticité et le statut de mathématiques chez Hume', *Revue de Métaphysique et de Morale* 89, 1984, pp. 453–64

Steiner, M. 'Kant's misrepresentations of Hume's philosophy of mathematics in the *Prolegomena*', *Hume Studies* 13, 1987, pp. 400–17

Stephen, L. *History of the English Thought in the Eighteenth Century*, 3rd edition, London: Smith, Elder, & Co., 1902 (1st edn 1876)

Stewart, M. A. (ed.), *Studies in the Philosophy of the Scottish Enlightenment*, Oxford: Clarendon Press, 1990

Strawson, G. *The Secret Connexion. Causation, Realism, and David Hume*, Oxford: Clarendon Press, 1989

Stroud, B. *Hume*, London: Routledge & Kegan Paul, 1977

Suchting, W. A. 'Hume on necessary truth', *Dialogue* 3, 1966–7, pp. 47–60

Thijssen, J. M. M. H. 'David Hume and John Keill and the structure of continua', *Journal of the History of Ideas* 53, 1992, pp. 271–86

Thomas, M. 'Relative ideas rejected', *Hume Studies* 8, 1982, pp. 149–57

Tweyman, S. 'Hume on separating the inseparable', in W. Todd (ed.), *Hume and the Enlightenment*, Edinburgh University Press, 1974, pp. 30–42

'Some reflexions on Hume on existence', *Hume Studies* 18, 1992, pp. 137–49

Van Sant, A. J. *Eighteenth-Century Sensibility and the Novel. The Senses in Social Context*, Cambridge University Press, 1993

Walsh, W. H. 'Hume's concept of truth', in G. Vesey (ed.), *Reason and Reality*, Royal Institute of Philosophy Lectures, vol. 5, 1970–1971, London: Macmillan, 1972, pp. 99–116

Waxman, W. *Hume's Theory of Consciousness*, Cambridge University Press, 1994

'The psychologistic foundations of Hume's critique of mathematical philosophy', *Hume Studies* 22, 1996, pp. 123–67

Wilson, W. 'Is Hume a sceptic with regard to the senses?', *Journal of the History of Philosophy* 27 (1), 1989, pp. 49–73

Winkler, K. P. 'The new Hume', *The Philosophical Review* 100 (4), 1991, pp. 541–79

Wolff, R. P. 'Hume's theory of mental activity', in V. C. Chappell (ed.), *Hume*, London: Macmillan, 1966, pp. 99–128 (First published in *The Philosophical Review* 69, 1960)

Wright, J. *The Sceptical Realism of David Hume*, Manchester University Press, 1983

Yolton, J. W. *Perceptual Acquaintance from Descartes to Reid*, Oxford: Blackwell, 1984

A Locke Dictionary, Oxford: Blackwell, 1993

Essay-review of E. Craig, *The Mind of God and the Works of Man*, in M. Stewart (ed.), *Studies in the Philosophy of the Scottish Enlightenment*, Oxford: Clarendon Press, 1990, pp. 306–11

'As in a looking glass: perceptual acquaintance in eighteenth-century Britain', *Journal of the History of the Ideas* 40 (2), 1979, pp. 207–34

'Hume's "Abstract" in the *Bibliothèque Raisonnée*', *Journal of the History of Ideas* 40, 1979, pp. 157–8

'Hume's ideas', *Hume Studies* 6 (1), 1980, pp. 1–25

F. Zabeeh, *Hume, Precursor of Modern Empiricism: An Analysis of His Opinions on Meaning, Metaphysics, Logic and Mathematics*, The Hague: Martin Nijhoff, 1960 (2nd edn 1973)

'Hume on pure and applied geometry', *Ratio* 6, 1964, pp. 185–191

Index

Indexer's note: David Hume's writings are indexed under titles only. The works of other authors are indexed under the author.

Lightning Source UK Ltd.
Milton Keynes UK
UKOW04f0306221217
314870UK00001B/113/P